The Power of Pattern
Patterning in the Early Years

Alison Borthwick, Sue Gifford, Helen Thouless

ATM
Association of Teachers of Mathematics

Contents

Chapter 1 Introduction: the power of patterning — 3

Chapter 2 Progression in repeating patterns: teaching and learning trajectories — 7

Chapter 3 Border patterns and problem solving — 13

Chapter 4 Spatial patterns and subitising — 21

Chapter 5 Growing patterns — 31

Chapter 6 How to teach pattern — 38

Chapter 7 Introducing pattern to nursery children with autism — 45

Chapter 8 The children's learning — 52

Chapter 9 Reflections on teaching — 60

References — 67

Acknowledgements — 69

Introduction: the power of patterning

Part of the appeal of patterns is that they are emotionally satisfying and often beautiful. They occur in a wide range of contexts, including daily life and the environment, as well as the art and music of different cultures. Teaching patterning can therefore build on teachers' and children's individual interests and strengths, such as creative activities or language, and within contexts as diverse as drumming or forest school. This is important, because when we as teachers enjoy teaching mathematics and are intrigued by it, we are more likely to pass on our enthusiasm to children.

Recent research suggests that young children's patterning, rather than their number understanding, predicts their later maths learning (Rittle-Johnson, Fyfe, Hofer, & Farran, 2017). Individual children differ consistently in recognising patterns in a wide range of contexts, including repeating sequences and arrangements of dots, according to Australian researchers Mulligan & Mitchelmore (2009). Fortunately, pattern awareness can be taught, and this improves young children's number understanding, as their colleague Papic found (2011). Disadvantaged and low achieving children benefit particularly, according to Kidd et al. (2014). Sue and Helen were excited by this research and wanted to try it out with other teachers and children – and so the project described in this book began.

What is pattern awareness and patterning?

Pattern is defined by Mulligan and Mitchelmore (2009, p. 34) as *any predictable regularity, usually involving numerical, spatial or logical relationships*. In the early years, the most common kinds are:

- **repeating patterns:** sequences of things, images or actions

- **spatial patterns:** shapes with equal sides or angles and regular arrangements of items, spaced equally or according to a rule

- **growing patterns:** e.g. staircases with equal steps, or tiled squares

While we distinguish between these three types of patterns, there is often overlap between them. When looking at a child's finished construction, it may not be clear whether they were constructing using a repeating or a spatial pattern; only if you saw how they enacted the construction could you tell what type of pattern they had in mind.

It is four year olds' *understanding* of repeating patterns, and not merely continuing them, which predicts their mathematical attainment at age 11, according to Rittle-Johnson et al., (2017, p. 7). They asked children to identify the smallest part that repeats in a pattern or to make the same kind of pattern with different objects (e.g. for a *red– yellow–yellow* pattern, to make the same kind of *ABB* pattern using green and blue blocks). However, as Threlfall pointed out in 1999, children can construct alternating *AB* patterns mechanically, whereas mathematical thinking is required to analyse the pattern and identify the unit of repeat. Papic et al. (2011) found that children who learned about repeating and spatial patterns in pre-school out-performed their peers at both pattern spotting and number in general when they got to school.

It seems that analysing visual patterns is an important skill, but it is not clear how exactly pattern awareness helps children mathematically. It may be that identifying the same pattern structure, like *ABB,* in different contexts encourages a focus on abstract relationships rather than visual appearances, which is a significant development in young children's mathematical thinking. Sarama and Clements (2009) suggest that patterning involves early algebraic thinking in generalising relationships to different contexts. This is supported by the findings from Lee et al. (2011) that pattern recognition predicts algebraic proficiency. It may be that awareness of spatial relations and spatial visualisation is important, as spatial reasoning is also linked to mathematics achievement (Rittle-Johnson et al., 2017). A habit of looking for patterns is useful: for instance, spotting the repeating sequence of one to nine when counting above 20, or realising that 3 + 4 has the same total as 4 + 3, reduces memory load and helps understanding. Papic et al. (2011) point out that identifying units of repeat involves unitising, or treating groups as countable items, which helps with multiplication. Whatever the cognitive link, it is noteworthy that the key predictor for young children is identifying the unit of repeat with practical patterns of objects and does not involve abstract numbers.

A nursery child attempting a repeating pattern with leaves and twigs

The Power of Pattern: Patterning in the Early Years

In England, pattern is part of the early years curriculum: when we began the project, statutory assessment required five year olds to *recognise, create and describe patterns* (DfE, 2017) and guidance suggested that children use *familiar objects and common shapes to create and recreate patterns (*Early Education, 2012). There was no suggestion of children identifying pattern units and no hint that 'recreating patterns' might involve generalising pattern structures. The exemplification for the Early Learning Goals (STA, 2014) for five year olds consisted mostly of alternating colour patterns, with teachers' annotations describing them as such. In the English national curriculum (DfE, 2013) pattern is only mentioned for year one in the non-statutory guidance, recommending that children *recognise and create repeating patterns*, with no further support or indication of progression. However, while we worked on the project, a new Goal of *Numerical Patterns* was proposed (DfE, 2020). While this acknowledged the emerging importance of patterning, we were disappointed that it was not based on the research for this age group, which had found practical patterning, with an emphasis on identifying and generalizing the unit of repeat, to be significant.

The Australian research offered more varied and appropriate patterning activities: we were keen to try these out and see whether they made a difference to children's mathematics. Sue and Helen recruited a small group of enthusiastic teachers of three to five year olds, from diverse London schools. We suggested a sequence of patterning activities and encouraged teachers to adapt these (knowing they would do this anyway!) Teachers and children developed activities in creative ways we did not anticipate and to a level which far exceeded our expectations. We presented the project to various groups of teachers who were keen to try out the activities with their children and to explore developments. Alison was similarly enthused and, two years after we began, formed an East Anglian professional development group, who followed the path of assessment and activities developed by the London group.

Assessing children's pattern awareness

We began each year by assessing a small group in each class, identified by teachers as 'number novices', 'apprentices' and 'experts' and asked these children to do some number and pattern tasks. (We used these terms to imply that children lacked experience in number, rather than numerical 'ability'.) This enabled us to compare their patterning with their number understanding. We were particularly keen to track the impact of the project on lower attaining 'novices', but also wanted to know how 'typical' and higher achieving children might respond. The children were offered four pattern tasks adapted from Mulligan and Mitchelmore (2015), which they did again at the end of the year. These included repeating, spatial and growing patterns.

Linear repeating patterns

Copy a 1-unit ABC pattern; then copy and extend a 3-unit ABC pattern

Border repeating patterns

Create a repeating pattern in a 12 or 14-space border.

Spatial patterns: subitising

When quickly shown first 5 and then 8 dot patterns, say how many dots there are.

Spatial and growing patterns: dotty triangle

*Copy a triangular pattern with 3 dots and then 6 dots (spatial pattern)
then extend the pattern: 'What do you think comes next?' (growing pattern)*

The varied responses of the focus children gave us clues as to the likely interests and difficulties of the broader group. We assigned levels of 'awareness of mathematical pattern and structure', which ranged from 1 (not copying at all) to 5 (extending patterns). These helped to indicate the range within each class, and gave the possibility of measuring progress across the year.

Overview of this book

As a group we developed progressions of activities for the different kinds of patterns, for nursery (three and four year olds) and reception (four and five year olds). Our proposed learning trajectory for repeating patterns is included in chapter two and extended into cyclical patterns in chapter three. Spatial patterns and subitising activities are in chapter four, including children's surprising responses to the dotty triangle task. These relate to the development of growing pattern activities, in chapter five. In chapter six we summarise our findings about effective ways to teach patterning. Children with special needs were included in the project, and their responses gave valuable insights into their thinking: a case study of a teacher of children with autism is included in chapter seven. The effect of the pattern project on children and teachers, including those in the subsequent East Anglian cohorts, are reported in chapters eight and nine. There are further suggestions of activities to try at the end of most of the chapters.

2 Progression in repeating patterns: teaching and learning trajectories

Sean had made a pattern with the little coloured bears, using a *red, green, green, yellow* unit of repeat. Simon, his reception teacher, told him it was an ABBC pattern. Sean said: *So you mean it could be 'dog, cat, cat, sheep'?* Impressively for a five year old, he not only understood the idea of representing the pattern structure with letters, but generalised this to a different context of varied animals. Sean was evidently well along the early years patterning trajectory.

Repeating patterns are perhaps the most common kind that we teach in the early years. These are usually linear patterns with a core unit, like the colours *blue, red*, which repeats, such as *blue, red, blue, red, blue, red*... They may be very simple with alternating colours like this, or more complex, prompting higher levels of analysis and generalisation, as Sean demonstrated. It is this patterning ability of four year olds to identify the unit of repeat and replicate a pattern structure using different items which is *a unique predictor of later mathematics achievement, over and above other math and non-math skills*, according to Rittle-Johnson et al. (2016, p. 12). It seems that patterning can help children to become pattern spotters and therefore provides an important access route to mathematics for young children.

Repeating patterns with sticks, shells, towers and pegs

In our project for three to five year olds we based activities on the sequence suggested by Papic et al. (2011), omitting some, such as drawing patterns, which we thought introduced an extra layer of difficulty. As teachers presented pattern activities to the children and observed their responses, they discovered progressive levels of challenge and so we developed our own flexible programme. We suggested using a variety of objects, such as coloured bricks, small toys, or twigs and leaves. Teachers included other modes like printing and using interactive whiteboards.

Papic et al. (2011, p. 253) encouraged us to go beyond alternating or ABC patterns, in order to avoid the misconception: *That's not a pattern... because it can't have two of the same colour next to each*

other. They suggested that the positive effects of patterning might be due to teachers encouraging children to compare their copy of a pattern with the original and to look closely, asking: *What's the same and what's different?* We began to realise that even copying was not the simple closed activity we had previously thought. Contrary to expectations, children found it easier to continue a pattern by chanting a sequence (e.g. *red, blue, green, red, blue, green*..) than to copy it by the laborious process of looking, finding, placing and checking each item. (However with action patterns the reverse was true, as children could copy movements, but found it difficult to hold the pattern in their heads in order to continue it.) Novice patterners could only hold on to a pattern for two units, 'losing' it on the third one; in contrast, pattern experts could pick up a whole unit at a time to continue a pattern.

Disagreement about what goes next prompts discussion

Slight changes could alter the level of challenge: for instance, children found it easy to recreate a pattern using the same colours but with different items, and harder to make the same kind of pattern using different colours.

Recreating a 'red yellow' cylinder pattern with bears is easier than trying to make the same AB pattern with blue and green cylinders

We realised that changing colours requires children to abstract the pattern structure, for instance thinking of an *AB* unit as requiring two different items, instead of just copying the colours. However, if we asked children to change just one colour, (e.g. replacing the yellow with a different colour) many could do this. One teacher called this 'editing' a pattern.

Copying then 'editing' an ABBC pattern, by changing some of the colours.

Spotting a deliberate mistake provided further challenge: we found this was easier when an extra object was added but harder when one was removed; fixing the error was harder still. In this way, we developed a rich progression of daily pattern challenges, with varied resources and contexts.

The same progression, from continuing, through generalising to fixing mistakes, can be repeated with increasingly complex pattern structures. *ABC* patterns require more thought than *AB* patterns, which children may just make by using left and right hands alternately, or by following a rhythmic chant. Younger children found it hard to progress beyond *AB* patterns, but it was not necessarily the length of pattern unit which mattered. Patterns are easier when all the elements are different, as with *ABC* or *ABCD* patterns, rather than when elements are repeated, as with *ABB* or *ABBC*. However patterns also seem to be easier if all the elements repeat, so *AABB* is easier than *ABB*. At this level of complexity, children seem to readily use letters to label the pattern units, and to generalise patterns to a range of contexts: Fyfe et al. (2015) found that teaching four-year-olds to use abstract labels helped them to do this.

Patterns with a long unit of repeat (like ABCDEF) can be simple to make

The progression presented here is a suggestion based on our experience: levels of difficulty depend on contexts and children's interpretations, so that in practice these stages may overlap. Within each kind of activity, such as error spotting, there are different levels of challenge and many possible variations. Knowing what to look out for in children's responses helps in planning further activities and challenges.

Progression in patterning: challenges and responses

The following apply to any pattern structure e.g. AB, ABC, ABB etc.

Continuing and copying a pattern:

- continuing an AB pattern by using alternate hands
- repeating the verbal pattern while physically continuing the pattern
- copying or continuing for two units, but losing it on the third unit

- copying by matching cubes one at a time
- using complete units e.g. picking up a red and a blue cube together when copying or continuing
- continuing a pattern which ends (or begins) with an incomplete unit

Explaining the pattern and identifying the unit of repeat:

- saying a continuous string: e.g. *It goes, red, yellow, red, yellow, red yellow....*
- saying the unit e.g. *It's a red yellow pattern,* or practically separating out the unit from a linear pattern
- abstracting and symbolising: *It's an ABB pattern.*

Generalising - copying a structure with different items

e.g. making your own *ABB* pattern:

- change the objects, using the same colours e.g. bears instead of cubes
- edit: change one item e.g. twig, leaf, leaf: replace twigs with acorns or shells; or change one colour e.g. replace blue with green in a red, blue, blue pattern
- abstract the pattern: e.g. change the colours of a cube pattern
- generalise: make an *ABB* pattern with anything in the room or outdoors
- 'translate' into different modes: e.g. using sounds or symbols
- create your own pattern structure: e.g. *'make any pattern so long as its not AB, with the bricks',* or with chosen materials

Spotting an error and fixing it:

- spotting an addition e.g. *ABCABCABBC*
- spotting an omission e.g. *ABCABCAC*
- spotting a swopped item e.g. *ABCABCBAC*
- correcting the mistake, by starting from the beginning again
- quickly correcting the mistake, explaining why it was wrong
- creating a pattern with a mistake for someone else to spot

Initially, we suggested that action and sound patterns came before object patterns, thinking them easier and more suitable for younger children. In fact, we found they were harder to copy and continue: on reflection, this is obvious, since the unit of repeat cannot be seen and keeps disappearing.

Children making action pattern sequences

Making a pattern sequence with musical instruments

At a higher level, action patterns also offer children the creative challenge of representing them with pictures or symbols, which may then be read and acted out by other children. This involves generalising by 'translating' patterns between active, pictorial and symbolic modes.

Children's pictorial symbols for action and music patterns

Music is another mode which offers opportunities for pattern, by simply alternating musical instruments, or creating more complex patterns involving rhythm and pitch. Movement and music can link with a range of cultures, including dance or drumming patterns from around the world. What is clear from children's responses to these activities is the way patterning can involve problem solving in a range of contexts, with children using a variety of strategies to continue patterns and to identify errors. Children also show creative solutions by making their own patterns in different modes and recording these. In the next chapter we show how young children can develop high levels of reasoning with cyclical patterns.

Examples of children's patterns – indoors and out

12   The Power of Pattern: Patterning in the Early Years

Border patterns and problem solving

Two boys in Karen Moses' reception class had succeeded in making a continuous border pattern with pom-poms around a rectangular mirror. This involved a repeating unit of three white pom-poms, two yellow and one blue, resulting in a very wavy border around the rectangle. When Sue asked them if they could make it fit better, one of the boys said: *I might change the pattern*. He then removed one of the yellow pom-poms from each unit of repeat. This seemed an impressive example of patterning by a five-year-old: instead of simply removing one of the repeating units he had altered the structure of the unit. Asked how they would tell someone else to make the pattern, the boy said, *White, white, white- three whites* and the other finished ...*Blue, yellow and start over and over*.

Pom-pom pattern

This chapter tells the story of a lesson, focusing on cyclical repeating patterns. Karen's reception class had progressed from linear patterns to making border patterns around different shapes. Creating a border pattern was one of our initial assessment tasks (based on Mulligan et al., 2015), which involved making a pattern with cubes to fit a border of 14 squares. While children could do this by using a simple *AB* unit to make a continuous repeating pattern, we found this task posed several challenges:

- **fine motor control:** some children found it tricky to place one block on each square.

Fine motor challenge

- **turning the corner:** some children started an *AB* pattern, but 'lost' the pattern as they turned the corner. We realised that, in order to rotate the pattern round the corner, you need to keep the pattern structure in mind, which means you have to identify it first. Children who made *AB* patterns as a mechanical activity, for instance using alternate hands to place two colours, found this demanding.

'Losing' the pattern at the corner  *AABB pattern unit not 'working'*

- **continuity:** some children had difficulty identifying whether a pattern 'worked', or would *go on round and round*, and were happy so long as all the squares were filled.
- **finding pattern units which 'work':** some children used more complex units, such as *ABC* or *AAB* and seemed puzzled when these would not 'work' with a border of 14 squares. The challenge is to find the size of units which can be repeated to fit into a fixed number of squares. (Borders of 12 offer more possibilities for this, since units with two, three, four or six elements will 'work', because they are factors of 12.)

Border patterns therefore offer rich challenges with different levels of difficulty for young children, from hand control to reasoning involving factors of numbers. Having observed children's various difficulties, one teacher hit on the brilliant idea of making cyclical patterns on paper plates: this removed the difficulties of hand control, turning corners and fitting the pattern unit into a fixed number of spaces. It focused attention on the idea of making a pattern which *carries on round and round,* and allowed children to adjust spacing to make the pattern 'work'. However, some children then wanted to fill in the whole plate, which diverted their focus away from the border pattern. Another teacher found large shapes to make borders around, including circular and rectangular mirrors and glittery foam shapes. These offered corner-turning challenges, without presenting a fixed number of spaces. Another teacher then produced lines of squares, so children could make linear patterns focusing on fitting the unit into the number of spaces, without turning a corner. Another produced circular borders with a number of squares and a series of border templates with different numbers of squares. We now had a whole progression of cyclical pattern challenges, which could all involve *AB* or more complex structures.

Progression in border patterns

Does the pattern work?

Paper plate pattern

Can you make a pattern going round the corners?

Glittery foam shape and yoghurt bottle tops

Pompoms around a wooden block

Can you make a pattern which fits the number of spaces?

I counted the spaces, so I knew I could fit it in twice

Which pattern units will fit into these borders?

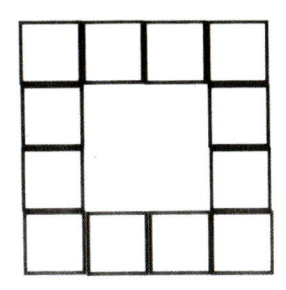

A reasoning lesson

The boys with the pom-poms in Karen's reception class had been introduced to border patterns in whole class sessions. Typically Karen would begin a lesson with the class sitting around the edges of the carpet looking at a pattern she had made before they came in. One day there was a sparkly foam triangle bordered with orange and green yoghurt bottle tops (donated by parents). The *AB* pattern had a mistake of two oranges together, giving rise to the following exchange:

Karen: *Have I got it right?*

Children chorus: *No!*

Karen: *What's wrong?*

Child: *One green, one orange, one green, two oranges.*

AB pattern with a mistake of two oranges together.

When Karen invited the child to 'fix it', she came out and, instead of removing the extra orange, proceeded to change the *AB* pattern to *AABB*. However, she got stuck going round the corner and said: *I have to think*, and then: *I need some help*. Karen suggested she could *phone a friend*, so she chose another girl to come out and help her. The two girls collaborated to finish the pattern, with the friend turning the corners. The following exchange focused on the unit of repeat, reflecting their previous discussions.

Karen: *What's different about their pattern and my pattern?*

Children: *It's two oranges and two greens and yours was one orange and one green. It has two the same colour.*

AABB pattern with four orange tops at a corner

However, the pattern did not 'work', as there were four oranges meeting around a corner. Karen asked: *Does it work? Hands up who thinks Yes*. Lots of children voted *Yes*, presumably because the yoghurt tops met and joined up all the way round. She then said: *Hands up who thinks No,* and: *Tell the person next to you why you think Yes or No.*

Karen: *What's wrong?*

Children: *Orange next to orange. Four of them. There's four oranges together.*

When Karen asked, *What can we do?* one child suggested, *Take one orange and one green out*, and then came and did this, so the pattern structure returned to *AB*. Karen then said, *Everyone, thumbs up if it works* and then, *Shall we check it and see?* The class joined in while she checked, saying, *AB, AB, AB* all the way round.

Karen concluded the session by summarising their findings on a whiteboard: *So on this shape, does an AB pattern work?* When they agreed she wrote *AB* with a tick. She then asked: *Does an AABB?* and recorded their response as *AABB* with a cross. Finally, she asked: *What else might we try?* Following suggestions, she wrote *AAB* and *ABB* on the board, so it showed:

AB   ✔

AABB   ✘

AAB

ABB

Karen's challenge to the class was: *Can you make one of these work?* (pointing to *AAB* and *ABB*). *But not just a triangle*. (There was a range of other shapes to choose from, including rectangles.) *We don't have enough bottle tops: what are you going to use?* To which a child responded, *Everything*! Finally, Karen asked: *Do you think you can do this?* The class shouted: *Yes*! This rally was repeated several times, and then the children dispersed in twos and threes to make patterns with their chosen shapes and objects.

Challenge time

Once the whole class session had finished the children had free time, which was always referred to in class as 'challenge time'. The children knew this was their independent time, when they could choose what to do within the provision of the reception classroom, but the expectation was that they would keep learning and challenging themselves (encouraged by Karen's rallying cry earlier). Most children chose to keep working on border patterns, taking up the challenge to find different units of repeat to fit around a variety of shapes. This was an inclusive activity offering a variety of challenges at different levels: for example, some children still found it difficult to turn a corner and needed support to work out how to continue their pattern in a new direction.

How do you turn the corner?    *Does the pattern join up?*

This activity also offered rich options in using a range of shapes, objects and pattern structures which would 'work' and fit closely. One group of girls challenged themselves to make a pattern with coloured magnetic letters around a triangle. They investigated different pattern structures, saying: *We won't do AB because that's too easy. We can try three first.* Having finished making the pattern they said: *It's an ABC pattern, we tried it and we writed down that it works – we even did ticks and smiley faces!* The girls then realised that they had left spaces at the corners and added another unit to resolve this.

Having been involved in reasoning in the whole-class session, the children continued to use reasoning during their free-play. Some began to reason about units of repeat in terms of the number of elements in them. Dylan worked with his friends to continue a pattern around a rectangle. However, they didn't keep all of the bottle tops close to the shape, preferring to make the pattern 'work'. Once completed, Dylan commented: *It's not fair because we haven't made the pattern fit properly – look, we've left lots of spaces.* He then set about fixing this problem, which then meant that his pattern would not 'work'. Having found that an ABC pattern did not fit around the rectangle, Dylan reasoned: *We need to try a different pattern, ABC doesn't work. We can try AB first because that's two and it might work better.*

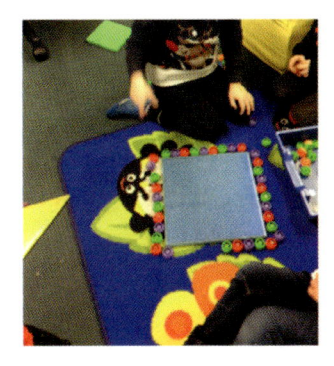

Trying different sized units    *A unit of four around a square*

Byron used the carpet session square and started to change the AB pattern into an AABB pattern because: *I want to see if an AABB pattern will fit around the square, AB did so two works but I don't know about four yet.* He built the pattern around the shape, lifted it at the end and checked whether his pattern worked. *It does work – look! AABB goes on and on all the way round.* One group of children worked together to create an *AABC* pattern and then separated the units of repeat out, explaining, *There's three units of repeat and there's three of us, we can put our names on one each!* In talking about the number of units, these children were beginning to unitise, treating each group as a single item, an idea which underpins multiplication.

Counting Units

Drawing a border pattern

Dylan continued with his investigations into pattern around shapes and decided that he did not need the physical shape because he could draw it himself. He then began to design *ABB* and *ABC* patterns which would fit around his shapes. Dylan was moving from making patterns practically to pictorial representations.

This session is impressive on several counts: there was serious engagement and reasoning involved in solving the problem of fixing pattern errors. This might be considered more typical of six year olds, but it shows what five year olds are capable of, given sufficient access to engaging practical problem solving activities like these, accompanied by appropriate questioning. This session also showed the effectiveness of a short whole class activity followed by opportunities to continue exploring ideas independently and over time. We found it notable that the children displayed many of the Characteristics of Effective Learning (Early Education, 2021) including:

- playing with what they know
- being willing to 'have a go'
- being involved and concentration
- making links and noticing patterns
- making predictions
- testing their ideas
- checking how well their activities are going
- changing strategy as needed.

Karen was clearly committed to building a mathematical learning community, with a safe risk-taking environment and an ethos of respect, as well as the expectation that reasoning is a necessary part of exploring mathematics. The children were genuinely intrigued by the enquiry and felt confident and comfortable in trying to solve problems, getting stuck and asking for help, offering analyses and suggesting solutions. They were expected to relish challenges, make their own choices, collaborate, and discuss. They also had the expectation that their teacher would be intrigued by their enquiries and celebrate their discoveries. Once these expectations had been set up, and revisited on a regular basis, the children were able to use, explore and expand their understanding of the mathematics of patterns.

Cyclical patterns therefore offer children opportunities to engage in the early algebraic thinking practices of *generalizing, representing, justifying, and reasoning with mathematical relationships* (Blanton et al., 2015, p.521). Patterning challenges like these, involving deliberate mistakes and prediction, practical exploration and safe risk-taking, can include all children, from those who find it hard to make a pattern turn a corner, to those who are developing unitising and reasoning with numbers.

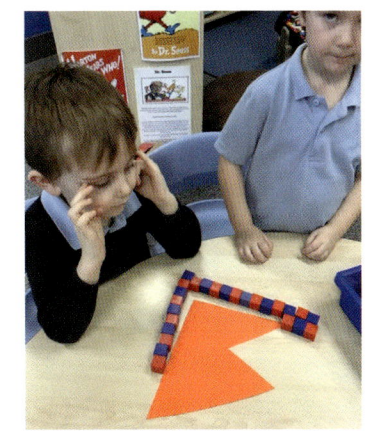

Problems with curves and corners

Activities to try

Challenge children to:

- make a pattern which continues 'on and on' around a border - and spot mistakes

Does it work?    *Can you fix it?*

How do you know/Why not? Can anyone fix it another way?

- fit a pattern into fixed number of spaces

Can you predict which pattern unit might fit?
How did you know it was/not going to fit?

- find different names for a unit of repeat in a continuous pattern

Depending where you start, what else could we call the pattern unit, instead of ABB? Could it be BBA? BAB? ABA?

Spatial patterns and subitising

What spatial patterns can you see in these pictures?

Spatial patterns

Spatial patterns involve items, some of which are the same, arranged in a regular manner (Papic & Mulligan, 2007) e.g. objects are equally spaced, make a regular shape or are placed according to a rule, such as being symmetrical. Some children spontaneously create patterns like this in their constructions or drawings: for example the boy in the picture above has tried to make a large hexagon out of smaller hexagons, completing it by placing the red and blue shapes symmetrically.

In the construction, the uprights in front are placed at regular intervals and there seems to be a rule about placing a smaller shape on top of them. There is a growing pattern of arches leading to the shape behind, which has both rotational and reflective symmetry.

There are many spatial patterns in nature. The petals of the pink campion are symmetrical and radiate at regular intervals around the pistil. The ferns are almost symmetrical and have a repetitive structure of serrated leaflets arranged almost opposite each other, growing in size.

One kind of spatial pattern which we focused on in the project were dot arrangements, as on dominos and dice, or in arrays or triangles.

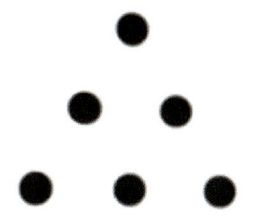

6-dot triangular pattern to copy   *A frequent interpretation of this pattern*

One of our assessment activities involved copying a spatial pattern of six dots arranged as a triangle (Papic & Mulligan, 2007). Whereas we saw this as a growing pattern and one of a series between a three-dot and a 10-dot triangle, we quickly realised that young children lacked experience of growing patterns and so were unlikely to see it like this (Thouless & Gifford, 2019). Some young children saw this pattern as isolated elements, either dots, triangles or rows, or they focused on the number of dots. Others saw a combination of these different elements, without seeing them as an integrated whole. Most children saw the dots and the triangle and so drew a triangular outline of dots, emphasising the visuo-spatial triangular nature of the pattern. This may be because young children tend to look for a 'contour' in an image (Sarama & Clements, 2009).

There are dots

There are six items

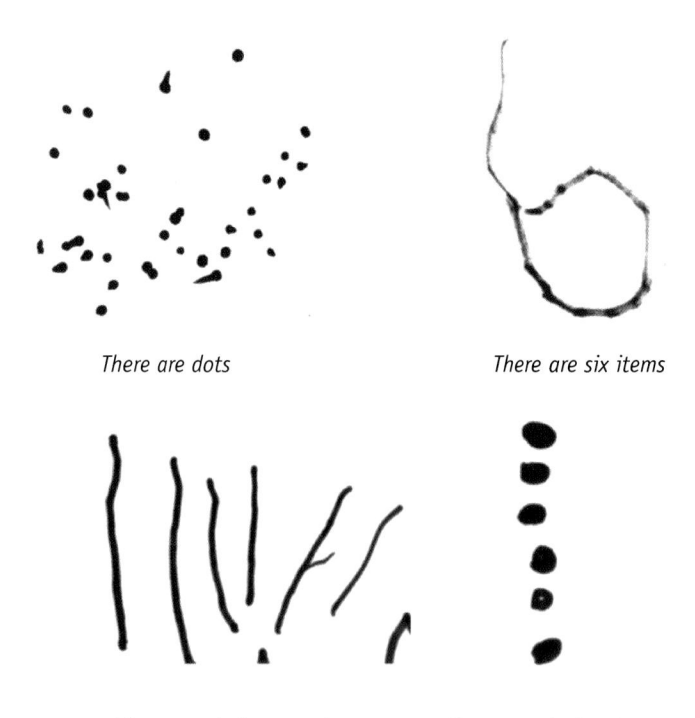

There are six items and a row.

There are six dots in a row

There are triangles, rows and sixness

Of those children who were able to copy the pattern accurately, most saw it as a whole triangle with equally spaced dots. As such, they were unable to extend the pattern, because in their minds the spatial structure was complete. One child who made this thinking clear to us was Sarah. Sarah first drew three dots at the corners of her shape, then an inverted v shape and finally put dots in the middle of each of the three sides of the triangle. She had efficiently analysed the organisation of the spatial structure and reproduced it in such a way that she maintained the equal spacing between the dots; however, the request to draw the next line in the growing pattern made no sense to her and so she did not do it.

Sarah emphasised the shape and spacing of the dotty triangle

Children's response to this task therefore gave us some clues about the challenging process young children may go through when trying to make sense of an arrangement of dots, in order to copy it.

Subitising

As part of our work on spatial patterns we focused on subitising. Subitising occurs when children can see a group of objects and immediately know how many there are: this is easier when the objects are placed in a regular arrangement, such as a spatial pattern. Subitising is an important skill, which helps children develop number sense (Clements, 1999).

There are two types of subitising: perceptual and conceptual. Perceptual subitising involves instant recognition of numerosity, which even very young children can do with one, two, or three objects (Sayers, 2015). Most adults can perceptually subitise up to four objects, although many of us are familiar with dice patterns so that we can also subitise five and six dots as long as they are in these formations. Most adults can also instantly recognise the number of fingers being held up on one hand.

Conceptual subitising involves the ability to quickly break a larger group of objects into smaller groups that can be perceptually subitised and then quickly added so that the whole is known (Sayers, Andrews, & Boistrup, 2016). With practice, older reception children can conceptually subitise up to ten objects, helping them to understand number composition and learn number bonds in a fun and practical way.

Mulligan and Mitchelmore (2016) have devised a learning sequence to develop subitising during the first three years of schooling, but we found that, before the children started on the learning sequence, they needed ample experiences of playing games to develop awareness of dice patterns and their fingers.

Dice games one to three

First, the children needed to be very secure with perceptual subitising up to three objects, so we created dice with only one, two, or three dots. With these simplified dice we made sure that the children played a variety of games, until they were absolutely comfortable with these three numbers. One teacher said: *When I put dice into every game it got the boys involved.*

One boy rolled the dice, recognised the number on the dice, and then made a tower containing that amount of cubes: Look, Miss, three is bigger than two.

Other games that we played were:

- roll the dice and get that many gems
- roll the dice and win a domino with that number on it
- roll the dice and do the action that number of times
- one child uncovers a pattern of dots and the other child decides how many dots.

Subitising fingers

My children liked to work with partners to show numbers under five on their fingers. They enjoyed discussing how to make the number and finding different ways to make the same number.

Finger gnosis, or finger awareness, is the knowledge of our fingers and being able to distinguish between them, for example in response to touch (Thouless, Hilton, & Webb, in press). Children use fingers for counting in two ways: to represent numbers (e.g. holding up three fingers to show their age) or to keep track of a count. Each of these functions is improved if the children can distinguish between their fingers.

Children who can show 'all at once' finger numbers are more successful with arithmetic (Marton & Neuman, 1990): one way of teaching children this important skill is for them to slowly 'grow' a number, by raising one finger at a time, then to 'show' a number by raising all fingers at once, and then to 'throw' a number, by instantly changing a fist into the number of raised fingers. Using two hands to show a number involves conceptual subitising, and can later emphasise number composition of 'five and some more' and other ways of making numbers.

- Simon says grow four, show four, throw four really fast!
- Show me five on two hands. Now can you show me seven?

Dice games one to six

Once the children were fluent working with numbers to three, we introduced the normal dice with dots up to six. They played all the games mentioned above and some new ones.

Roll the dice for the toy, then collect that number of pennies. Who has most?
(Griffiths et al, 2016)

Matching dice and domino patterns

One of the new games we played was *Ten Nice Things* (Skinner, 1997), or *Five Nice Things* for the younger children. This game is appropriate for between two and four children. Each child selects 10 nice things from a larger collection. The first child rolls a dotty dice, subitises the number and gives that many objects to the other child or the one on their right. This child then rolls the dice and gives that many to the next child. (As this is a game about giving, the winner is the first to give everything away.)

My children loved Ten Nice Things. They played it for ages.

Alternative patterns

Once the children were fluent in recognising dice patterns and fingers, they were ready to start developing conceptual subitizing. One way that the teachers helped the children to do this was by drawing alternative patterns for the numbers three to six, then asking the children to draw some and discussing what was the same and different between the arrangements for each of the numbers.

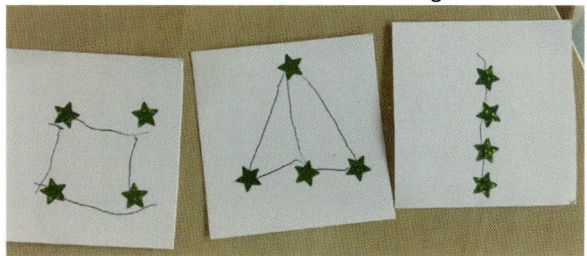

A nursery SEND teacher said, *I found that stickers worked really well. All the children were interested. I used the language, 'Don't count. Just tell me how many.' Even young children could do simple patterns. They used a variety of methods: four corners, line with one off, line.*

When asked to draw six in another way, a reception girl said, "It's five in a line and then one more"

Then the children created dot patterns for the numbers seven through ten. Some children used patterns for fours, fives, and doubles, others added to the pattern for six, while others drew straight lines for all of the numbers. The children then found another way to represent the same number. Once they had created their dot patterns, they were asked which patterns they liked best and why. They tended to like their own patterns. If not, they liked the ones which look most organised but couldn't verbalise why. A question that we asked to help them think about this was: *If you showed your friend, which one would they get first?*

A reception girl was clearly breaking numbers into smaller parts to create new patterns as she drew 8 in two different ways.

When asked to draw ten, a reception boy said: Easy, it's five and five.

The structure of the ten frame can support children to make some alternative representations of the numbers five through ten. Its regular structure, with evenly spaced lines and spaces and the two rows of five, helps children to conceptually subitise. For example, *I see one row of four and another row of four, so there are eight dots* or *I see two less than ten, so it is eight.*

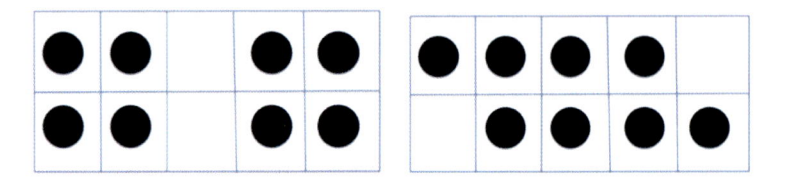

The number of the week is eight. Tell me quickly whether this shows eight.

Two ten frames can then be used to help the children work with teen numbers.

The regular structure of Numicon©, which is based on the ten frame, can also support children to develop their subitising skills. For example: *I see two even rows of four, so this is eight.*

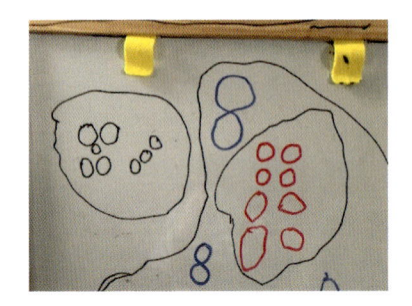

Children making Numicon© patterns by printing on playdough and drawing

One teacher used slides with the numbers one to ten in Numicon© patterns and showed the images too fast for the children to count:

> *I then asked them how many dots and how did they know. The children often used their hands to justify their answers, which is a generalisation from Numicon© to fingers. The children really enjoyed this activity.*

I can see four, four, and one; I can see five and four; eight and one.

The teacher noted that most of his reception children could identify the number when it was split into two groups, but that fewer children were successful when there were three groups. Sometimes the children made the pattern they saw and then discussed it. The children also matched Numicon to dice, which supported the children in transferring their knowledge of dice patterns to a different format.

We found other formations of dots that supported children with both perceptual and conceptual subitising.

One nursery teacher developed a Bingo game to support children's perceptual subitising. They had to find the square with the number of spots that matched the number called and then use a Bingo dabber to hit each dot e.g. when the teacher called six, the children found the square that had six dots and then hit each of the six dots with the dabber.

Several teachers mentioned that whenever they had a few spare minutes they would put a small number of magnetic buttons on a small magnetic board or tin tray and ask the children what they saw. Or they would invite the children to find different ways of rearranging the buttons e.g. to show five. Depending on the number of buttons this could involve perceptual or conceptual subitizing. The teachers found that using real objects like the buttons, rather than dots on a screen, was more effective with young children. This also seemed likely to support children's understanding of number conservation, because they know that the real buttons are still there and so the number cannot have changed.

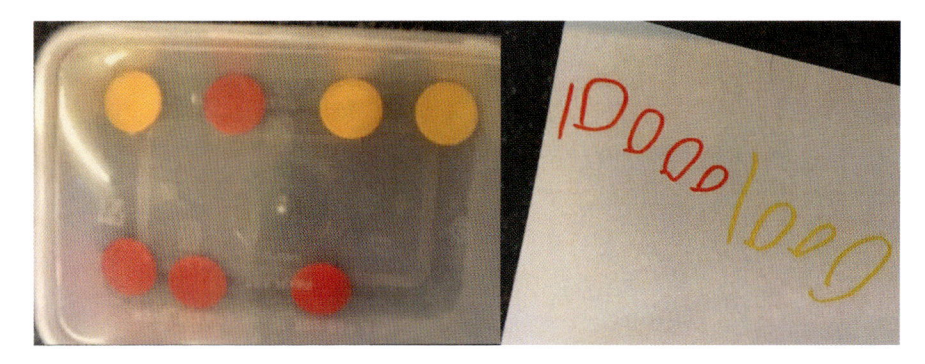

An activity for perceptual and conceptual subitising: shake a take-away container with seven two-sided counters for a partner to say how many red, how many yellow, and how many counters altogether, then record the result.

A reception teacher used dots on paper plates to give children practice at subitising. The plates were hung up in a bag by the board: the teacher would naturally pick up the paper plates slightly differently each time so the children got used to subitising the arrangements at different angles.

Games with two dice

The next step was to play games with two dice, using numbers to 12. Most games can be adapted in this way: *Ten Nice Things* can become *Fifteen Nice Things* so children work with numbers in the teens. Usually this step was not reached until the children were in reception, although there was one exceptional girl in the SEND nursery who reached this stage when she was three.

> She subitises up to ten with two dice. She can also put two Post-Its together [with dots on] and say that six plus three equals nine.

A reception teacher gave his children sheets of paper with different arrangements of dots to 12, and asked the children to show which groups they saw. Some children chose to write number sentences to match their picture.

Bears in a tent

One of our favourite number assessment tasks is *How many bears are in the tent?* As children become better at subitising they also become better at this task (Wang, 2020). This may be because they can visualise the number of bears in the tent, and then conceptually subitise what happens when more bears enter the tent.

> A reception girl, when asked how many bears are in the tent, answered with confidence:
> Of course, Miss, there're still four there.

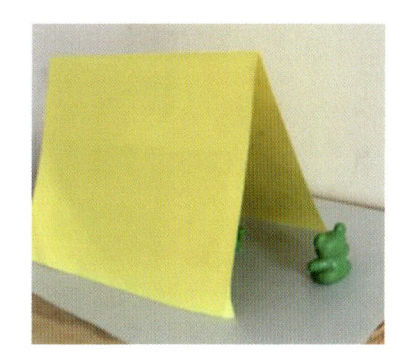

Three bears go into the tent. One more enters. How many bears in the tent?

Impact

Subitising is not easy, but increasing subitising skills improves children's ability to give a number of objects and to count objects accurately (Wang, 2020). As one reception teacher noted, this increase in counting skills may also lead to improved arithmetic skills:

> *A reception girl who had been struggling with arithmetic used the skills she had learned in conceptual subitising to solve 11 add four, because she moved one over to the four and then she knew the answer was 15. Another child saw 20 as four fives, five add five add five add five, which is the beginning stage of multiplicative reasoning.*

The impacts of subitising may not only be on mathematics, because those who are stronger at spotting patterns are also stronger at reading (Bragman & Hardy, 1982).

> *She is good at subitising two dice and can also recognise words through the pattern or shape of the words.*

Subitising can be practised in the context of games, which makes it more enjoyable for the children to engage in. They also enjoy in engaging in number talks, for which there is no one correct answer.

Activities to try

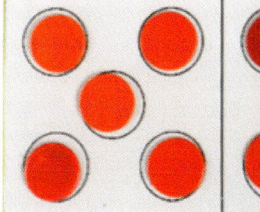

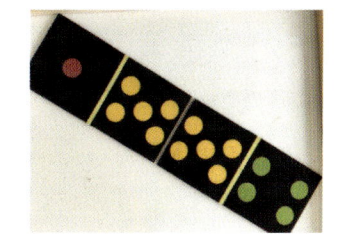

- subitise Hungarian number pictures
- play dominoes
- number talks: *Say what you see*
- play *Splat!* Swat an item matching the number thrown on a dice

- provide resources for children to create symmetrical patterns

- go on a hunt for things to subitise

5 Growing patterns

Growing patterns

Teacher: *Can you see that I have made the start of a pattern?*

How do I make it bigger? What's the next step in the pattern?

The child puts another line of three slightly more stretched out and comments on it being a Christmas tree and a pyramid.

Teacher: *How many at the top of the pyramid?*

Child: *It's got to be in layers. I'm going to put one, then two, then three, then four.*

The teacher then puts four cubes in a cross. The child extends the arms of the cross and then they discuss other ways it could be extended according to different rules, for instance growing downwards or sideways. Then they discuss: *What is a pattern?*

The teaching extract above is not a spontaneous or isolated incident. It is a product of the pattern culture that we have aimed to cultivate in our project. The teacher chose to engage a child with growing patterns who seemed very pattern aware.

At the beginning of the project, like Papic *et al.* (2011), we considered growing patterns to be more difficult, so we did not focus on these in our sessions. Young children do not readily recognise the *one more than* relationship between counting numbers: connecting the cardinal value of each stack with the *going up in ones* relationship helps children make the major leap in understanding that each consecutive number includes the previous number plus one (Gifford, 2014). This is the principle of hierarchical inclusion, which is involved in all growing patterns. Understanding this requires part-whole recognition, or simultaneously seeing the whole and its parts, which has been identified as a major milestone in young children's number knowledge. However, some teachers and children did not think that growing patterns were too advanced, particularly in one nursery class. Their work is described in this chapter.

Growing patterns using colour rods

Simon Lewis was working with a small group of nursery children, at the horseshoe table. This is something that he did twice a week with all the children and was an opportunity for him to investigate what his children knew and to observe their responses to a challenge. Simon had decided to introduce the children to colour rods: after a short numbering task Simon told the children that he was going to put the blocks in order from longest to shortest, but actually he put them in a staircase that went five, four, three, one, two. This was very surprising to Hassan and Lydia, and Hassan said: *Three doesn't come after one*. Simon asked: *Can you fix it?* and Hassan then repaired the error.

Simon had found that making deliberate mistakes in patterns was a fun way to engage children and to assess their understanding of patterns. Once the children had seen Simon's correct sequence of colour rods, they built their own. Lydia originally ordered the rods: five, four, two, one, three, but only corrected it when she saw Simon's model, not when she looked at the other children's correct models.

Colour rods staircase

Soon the children were no longer satisfied with building colour rod stairs just to five. One girl was interested in finding the rod that came after five. Once she found this rod, another girl said: *We will have to ask Simon to give us the number six*. The first girl responded: *I know where there is a six*, and she went and got the numeral six off another display. She then also matched the six rod with six one rods. Simon had given the children the space to explore the mathematics of the colour rods staircase and the children had the confidence to go beyond his expectations, which they continued to do, building on each other's ideas. When Adam put all the colour rods rods in a staircase to ten, the other children were fascinated and copied his staircase to ten. When one child made their staircase backwards, Simon helped to put it together with Adam's so that the staircases ascended and descended. They then put four staircases together so they were symmetrical both vertically and horizontally.

Reflective staircases

Later Adam was curious about what came after 10. He worked out that the next numbers would be 10 and one, then 10 and two, and then 10 and three. For the following number he put together 10 and six but adjusted the length of the six rod to continue the staircase shape, although that made bottom of the pattern no longer level. Adam was beginning to figure out the structure of the teen numbers, although he did not yet have the vocabulary to name these numbers: however he was confident that all the steps in the pattern had to be the same size.

Adam's staircase

Staircase patterns with different resources

Growing patterns using interlocking cubes

Soon the children were exploring staircase patterns with different resources. They were enthusiastic watchers of the *Numberblocks television* series and when they saw the episode with *Step Squad 15,* the class became very enthusiastic about making the *Step Squad* with a variety of materials.

Numberblocks Step Squad 15

In the photograph below the child seems to be focusing on the step-like visual image but does not recognise the organising principle of the stacks going up in ones. However, this is a good example of the emergent level of pattern and structure awareness.

An emerging growing pattern

Simon invited the children to stick coloured squares on a piece of paper to show the pattern: the children were very engaged by this activity, producing a variety of responses, apparently showing different levels of pattern awareness. Some children stuck the squares onto the paper randomly, some made number stacks, not necessarily in order, and some made a stepped staircase.

Marta made the staircase pattern as far as four and then made the dice pattern for five. We realised that the range of responses showed that children were paying attention to different aspects of the pattern, such as numbers and shapes, including alternative patterns. There were other resources that the children also explored. They made the *Step Squad* outdoors with blocks, bricks, *Numicon* and even cable reels.

Giant staircase pattern of cable reels

We are always encouraging any mathematics learning to go 'huge and outdoors' and pattern play is no exception! Some children used the idea of the step growing pattern to explore other areas of mathematics. On one occasion Yves built a staircase as tall as himself, by placing bricks long-ways up: in doing this he was exploring aspects of measurement and spatial awareness.

Is the tower as tall as me?

Dotty patterns

As we have mentioned previously, one of our pattern assessments involved asking children to copy a triangular growing pattern of dots. As described in chapter 4, most children saw this as a dotty triangle: however there were some exceptions.

Rosie was four and a half years old at the end of nursery. According to our number assessments, she was a 'number expert' who confidently recited numbers to 39 and counted out five jewels to give to a teddy. She had last seen the dotty triangle task nine months before, when she had responded by drawing a triangle-ish shape with a dotty outline, as many children did.

dotty triangle-ish shape

However, this time Rosie started by drawing a row of six dots, then another one above it, followed by a single dot above that.

Rosie's first drawing: two rows of six dots, plus one.

She then said: *What it looks like is three at the bottom, then two, then one*, and drew the triangular six-dot pattern.

Rosie's second drawing: correctly copied triangle.

She followed this immediately with a 10-dot triangle, drawing four dots on the bottom row, then three, then two, then one. When asked: *What were you thinking?* she replied: *This* and drew lines across the rows of three, two, and one dots, thereby obscuring the structure.

Rosie's third drawing: the 10-dot triangle, with lines drawn through the top three rows afterwards

The end result was a group of three drawings that would have been impossible to interpret had we not been observing her as she completed it.

Rosie's three drawings

Rosie appeared to have gone through several levels of pattern awareness in a couple of minutes. Her first version recognised several independent elements: number, dots, equal spacing and possibly rows (and perhaps she had added the top dot to give a somewhat triangular shape). She then described the structure and constructed the triangle by rows before generalising the structure to the next triangular number image. Rosie showed us that a four year old could recognise and continue a growing pattern. While few children spontaneously recognised the dotty triangle in this way, we realised this may have been because we had not shown them the growing process.

Implications

The nursery children's enthusiasm for staircase arrangements also seemed to indicate that growing patterns might be more accessible than we had assumed. Growing patterns can be valuable in providing children with access to number patterns and relationships. One example of this was the giant outdoor staircase pattern with cable reels. This stimulated an informal number talk session, with children standing on different sides of the pattern, noticing different things, such as the ascending and descending number sequences.

One child says: It goes, 1, 2, 3, 4, 5.

Another child, standing on the side says: I can see 5, 4, 3, 2, 1.

We have subsequently found that reception children enjoy making growing patterns with a variety of shapes and materials, and that these often have a numerical rule, such as going up in threes. This was a popular home learning activity during the pandemic. Like the reception's child's interest in creating rules for growing crosses at the beginning of this chapter, it seems that growing patterns have rich potential for developing reasoning, problem solving and generalising with young children.

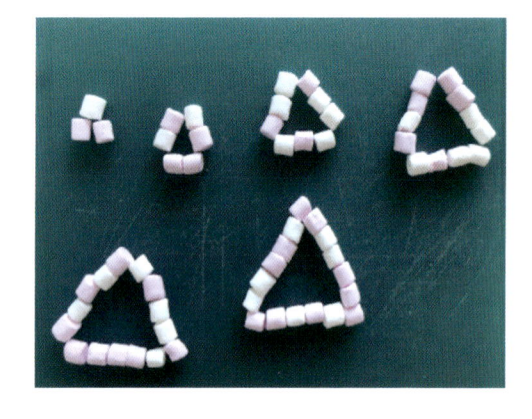

Activities to try

Challenge children to:

- create a staircase pattern for children to continue.

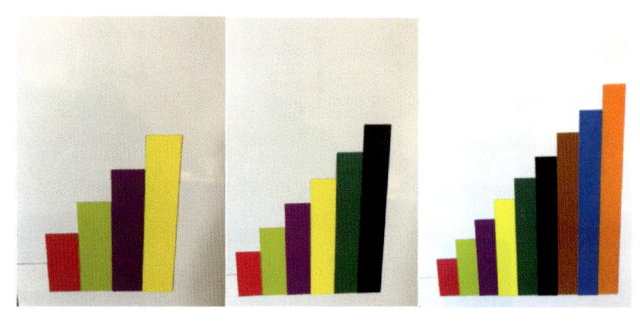

What comes next?

36 The Power of Pattern: Patterning in the Early Years

- grow some bigger shapes out of little shapes.

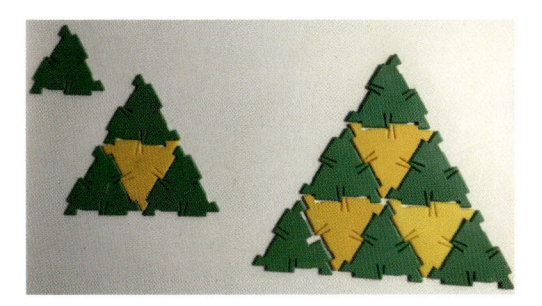

What is the rule for growing the next one?

- start with four or five tiles arranged in a T or L shape:

Make up a rule for how this grows - what comes next?

- Use a variety of objects to make a growing pattern.

A fruit triangle

- Spot some natural growing patterns.

What pattern can you see here?

6 How to Teach Pattern

This chapter summarises what we found out about successful ways of teaching pattern awareness in the early years, focusing mainly on repeating patterns and subitising. A variety of approaches were adopted, some of which were teacher-led and some more child-initiated, depending partly on teachers' current practice and the ethos of the school. Many teachers found that the most effective strategy was to have a short introduction with the class or a small group and then invite children to develop these ideas with selected materials indoors and out. Teachers would then observe what the children did and discuss their patterns with them individually. A more child-led approach was to observe children's play, then offer support to extend pattern ideas, perhaps in parallel play. In both cases, teachers increasingly found they were fascinated by the children's responses: *I just introduce the idea and then see what happens.*

Several teachers incorporated pattern into their daily 'maths meetings', which usually involved a range of mathematical topics and routines, such as calendars and counting. Whereas previously teachers had tended to teach pattern for one or two weeks a term, they realised that there were so many possible developments and applications, that they could easily provide brief daily pattern sessions. These included subitising and discussing spatial patterns in 'number talks', as described in chapter 4. Some teachers included whole class and small group sessions:

> *What worked well was maths meetings, then repeating in a small group for the non-verbal and quieter children, it allowed them to get more involved.*

Teachers found that older children could sustain whole class discussions for longer periods of time and engage in problem-solving sessions as described in chapter 3.

An introductory pattern session

The teacher starts a repeating pattern on the mat with the class or group sitting around:

> *I'm going to make a pattern: watch closely (no shouting out!)*
> Starts a block pattern: red, green, red, green, red, then pauses.
> *I wonder if you can read my mind - what do you think I am going to put next? Talk to the person next to you.*

Lots of children talking.

> *Now see if you were right.* Continues pattern for several repeats.
> I heard people saying, red, green, red, green ... Can you tell your partner, what is the unit of repeat - what is the smallest bit which happens again and again and again?

Children talking.

Who would like to pick up the unit of repeat to carry on my pattern- Lyra, can you come and show us?
Lyra comes out and picks up a red and green pair of blocks.
That's it – that's the bit I can use to carry on the pattern. So the unit of repeat is red, green and I also heard people saying AB. It's an AB pattern.

T: Now, have a look at another pattern I am going to do. I wonder if you can read my mind this time.
Puts down red, red, green, red, red, green blocks.
What is the same and what is different about this pattern and the one I just did?
Talk to the person next to you, what can you see?

Children talking.

T: Mikey said he can see red, red, green, red, red, green. Who can tell me something else about this pattern?
C: There's two reds.
T: There's two reds. So what is the unit being repeated? What unit can I pick up and put down here?
C: Red, red, green
C: There's three.
C: AAB
T: So the unit of repeat is red, red, green, AAB. It has three things.
C: Two the same colour.

T: I wonder if you can go and make a pattern like this red, red, green one, an AAB pattern? But see if you can make yours with different colours or different things, like the twigs and leaves. Or you can choose to make a different pattern.

The children then go off to make patterns independently or in twos and threes. In most classes, patterning is one of many activities available and children can make patterns at any time during the day, with pattern stations provided indoors and out.

Teachers would vary the introductory activity daily, with progressive challenges as described in chapter 2. They might vary the items used, moving away from colours to different objects, shapes, or sizes. One teacher set himself the challenge of using units of repeat or pattern rules that the children would find hard to identify, saying: *I'm going to see if I can trick you, can you guess my rule?* Increasingly he found that he could not outwit the children and the children delighted in devising their own pattern rules to 'trick' each other. The teacher would photograph their patterns and use them to begin the next day's session.

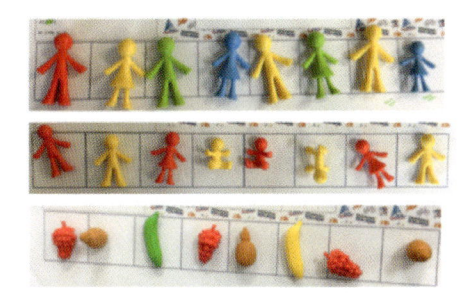

Children's patterns photographed and used as 'Guess my pattern' starters in the next day's session.

Some teachers encouraged children to spot deliberate mistakes by making a pattern in front of the children and saying: *Watch carefully: I'm not sure if I'm going to get this right.* For a harder challenge, teachers would lay out a pattern on the carpet before the children came into the room, and just ask:

What do you notice? Once children had spotted the mistake they were invited to explain how they knew and then to suggest different ways of fixing it and checking. In this way teachers modelled reasoning, encouraging children to explain to their partners and so developing discussion skills, as described in chapter 3. Sessions usually ended with a challenge to the whole class, which individuals or small groups could pick up in whatever way they chose.

Can you fit an ABB pattern around your shape?

Pattern in play

One nursery teacher would introduce patterns while engaged in parallel play with the children: sometimes he would describe the pattern explicitly: *I'm making a pattern with my tower... it goes this way, that way*. Or he might quietly make a pattern and see if the child noticed:

> *I was building with Lego beside a little boy. He was making a tower, so I made a tower and did mine as a pattern and he looked at mine and he was trying to copy.*

This observation provided valuable information about the child's pattern awareness. At other times he would observe a child doing something and think, *the others could do that too!* In this way he used observations to inspire his planning.

After observing a child's efforts to recreate Step squad 15 from Numberblocks on TV, the teacher encouraged other children to do the same, resulting in the class construction of the cable reel staircase

Integrating pattern

Some teachers integrated pattern throughout the curriculum.

> *Last year I did a section on one aspect of pattern at a time. This year it has been more integrated ... we had it all the time. Pattern has not been explicit but integrated into whatever we do.*

Patterning integrated into Forest School activities

Another approach was to integrate pattern into classroom routines:

Maths is more fun to teach. It's easy to bring it in 5-minute bursts. It's encouraged me to teach maths more through the routines of the day.

Just walking, taking them to lunch, I stamp in a rhythm: stamp, stamp, clap.

In all these approaches, the project teachers tended to use the same low threshold - high ceiling activities with all the children. This contrasted with the Australian approach (Papic et al., 2011), which involved assessing children's levels of pattern awareness and then teaching levelled activities to small differentiated groups.

Assessing pattern awareness

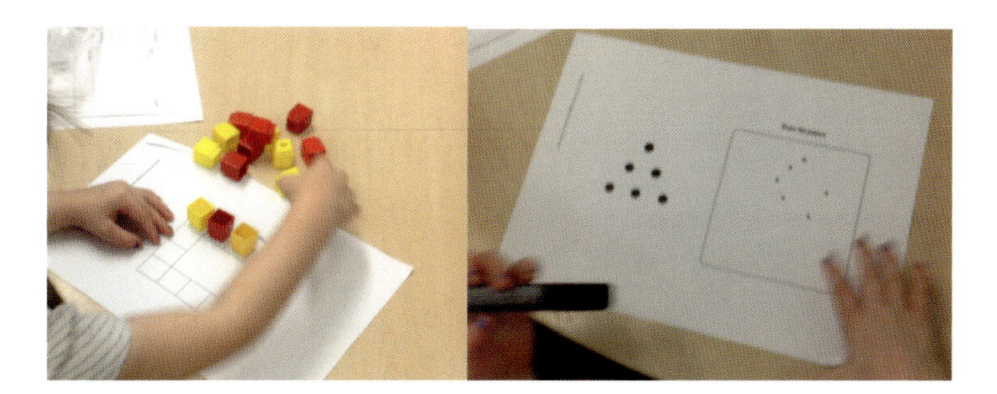

As described in Chapter 1, we tended to start and finish each year by assessing sample children's pattern awareness, in order to find out where to pitch the level of activities and to gauge progress. Using set assessment tasks was unusual for nursery teachers: however, one said she actually liked having the assessments, because it was good to see how the children responded and she liked to see their different ways of looking at things. Of course, the teachers also continuously engaged in assessment for learning, judging what the children understood and how to challenge them further, through careful observation and listening to children. The end-of-year assessments validated what the teachers already knew: most children had improved in their pattern awareness over the course of the year.

Classroom ethos

As the teachers worked on patterns, a distinctive classroom ethos developed, which was collaborative and social, with a high level of playfulness, challenge, and mathematical discussion. One reason for this was the teachers' use of differentiation by outcome, presenting challenges to the whole class which were accessible but extendable, as described previously. Patterning proved to be a highly inclusive activity: children are engaged by attractive manipulatives which can easily be rearranged - and the ceiling can be as high as they choose to make it.

Whole class discussion about patterns

Because all children were focused on the same challenges, it was expected that children work together. This allowed an ethos to develop where both the social and academic sides of learning emerged simultaneously. A deputy headteacher commented:

Co-worker working is really coming through. The children are actually wanting to share with their friends, they're learning to learn from their peers, but they are also sharing with the adults. But I think that social side of it, that need to communicate, that purpose for doing something, is quite evident.

Children working together and setting challenges

This social side of patterning was encouraged by the teachers. Sometimes when a child was stuck on a problem in whole class discussion, their teacher encouraged them to *phone a friend*. When a reception class was working on recording their action patterns with symbols, their teacher encouraged them to use an iPod to film a friend 'reading' and enacting the pattern.

The teachers playfully made mistakes to see if the children could spot them, which they found hilarious. This ethos of playful challenge did not remain top-down, merely coming from the teacher. Because the teachers modelled posing challenges, the children wanted to do this too. All of the teachers said that *one surprising aspect of the children's interactions* was their delight in setting each other challenges and tricking their friends. Interestingly, this aspect of patterning is rarely mentioned in the literature, a notable exception being Helenius (2018).

This emphasis on challenge led to a high level of mathematical reasoning within the classes. Children were expected to think mathematically and to express themselves mathematically. The teachers supported children in expressing their mathematical reasoning by asking questions such as:

- *What pattern could this be?*

- *Is it a pattern? Why? How do you know?*

An important aspect of their questioning was following up their initial questions with *Why?* The children were expected to explain their thinking and reasoning and to discuss this with partners, saying *because…* As identified in chapter 3, the classroom ethos fostered within these classrooms provided multiple opportunities to exemplify the Characteristics of Effective Learning (DfE 2021; Early Education, 2021), playing and exploring, active learning and creating and thinking critically.

Resources

Some teachers set up pattern stations as part of continuous provision. As well as a range of objects, these might include templates for the children to make linear or border patterns. Outdoor stations might include leaves and twigs, with long strips of paper and double-sided sticky tape to keep objects in place. Importantly, in these cost-conscious times, patterns can be made with any resources available: the important thing is that there are sufficient quantities to make satisfactory patterns. Teachers in a farming community used different sized potatoes, whereas those in an urban school used the colourful tops of baby yogurt pots, donated by parents.

Using available resources: potatoes

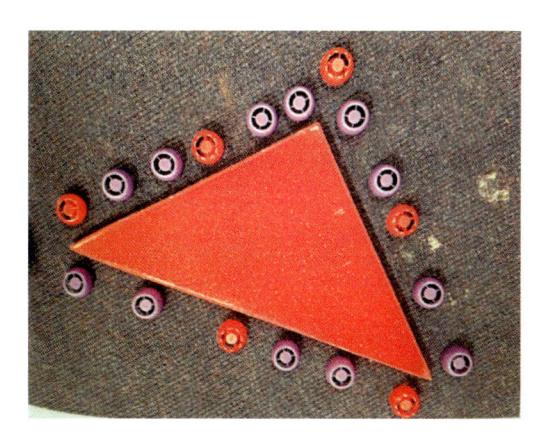

Resources provided by parents: yoghurt tops

Home learning

During the COVID-19 pandemic, the home became an important place for learning about patterns. A reception teacher created slides for both parents and children to support pattern learning at home.

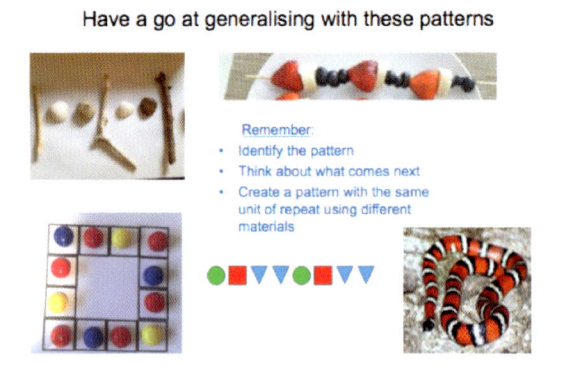

A slide for parents to support children's patterning. *A child working on pattern at home.*

Communities of practice for teachers

The most valuable part about our bi-monthly meetings was meeting with other teachers, exchanging ideas and seeing my work being valued. I also enjoyed having my teaching recognised.

We met together two to three times a term to exchange ideas and celebrate our successes. We found that getting a group together to discuss pattern helped us come up with new ideas and helped the children's learning progress. This was a very important part of our success as teachers, learners and explorers of pattern. We will discuss this aspect of our project more in chapter 8.

Teaching approaches to try

- set up pattern stations indoors and out, for children to further develop patterns introduced in whole class or group sessions

- model patterns which are not just about colours and with more complex structures

- take photos of children's patterns and use them as challenges the following day

- observe and listen to children when they are making patterns, to identify things to develop with the whole class

7 Introducing pattern to nursery children with autism

In this chapter we explore how a teacher of a special group of nursery children utilised and explored patterns in her classroom, drawing on a conversation between her and her former deputy head. While this case study shows how young children who have developmental challenges engage with ideas about pattern, it also suggests the broader significance of pattern awareness.

Mary's class

Mary Edgar is part of the pattern project and so has been part of the discussions and collaboration of the group since the beginning. While others were trying out pattern activities with their children Mary was also developing ideas about pattern, but in a different way. She teaches three and four year olds with a diagnosis of social communication disorder or autism, some of whom are non-verbal. When they first come to school, it is a potentially upsetting, unfamiliar environment, with many other children, even though Mary and her team work hard to ensure that they are in a calm unit, which is separate from the main nursery.

Rearranging the order of a visual timetable

Patterns in routines

The establishment of predictable routines is therefore crucial and visual timetables with pictorial symbols for activities are an essential way to communicate these routines with non-verbal children:

> ... Eventually they can manage their own timetable and remove items from the timetable as they finish an activity. Some children know the timetable so well that by the end of the year we can ask them to set up their visual timetable for the next day. I remember that Kenji knew the timetable so well that he could set up the timetable from the bottom up.

A predictable sequence of daily events helps children to manage transitions which might otherwise be upsetting: visual timetables make these explicit and under the children's control. Sharing this strategy with parents helped in the pandemic lockdown, when the children's daily pattern changed and they could no longer go to nursery. Mary realised that some parents were struggling to support their child. One parent phoned her, saying,

'I don't know what to do. You know, we can't go out and she's driving me mad as she keeps going to the door with her school bag on and looking at me. And I don't know how to stop that.'

So Mary suggested deciding with the child the night before what activities they were going to do, drawing pictures on sticky notes, sticking these on the wall and then next day going through the activities, taking them off when they had finished each one. The following week she phoned the parent and asked how it was going:

… And she said, 'We're fine. We've done everything that you said and she's really happy. She comes out every morning and does the first thing, and she's doing the same activities in the same order.' ..When the child came back to us after lockdown I couldn't believe the positive change in that little girl. ..After the lockdown and using the visual timetable, some parents said, 'I think we're closer as a family. We're able to do more things together. He's letting me read him books'. .

This improvement in the lives of families suggests the emotional significance for children of feeling in control of change: identifying a pattern enables them to predict events and prevent anxiety.

Patterns were embedded in all routines in the nursery, such as snack time:

The simple rules might be: we all sit to eat, we wait our turn, we request our choice and we go around the group in a particular order.

A child anticipates his turn at snack time.

Similarly, transitions were introduced with time warnings and countdowns:

At the end of each activity we will say, 'We will finish playing in three minutes' and we will set the timer, either an electronic timer or a sand timer. It becomes easier if towards the end of that time we also do a countdown. The children love joining in with that. We'll say,' Let's look at the timer. Oh, let's countdown. Ten, nine, eight, seven, six, five, down to zero'. And then as a class you will say, 'Play finished. It's time to tidy up'.

Routines were also accompanied by songs adapted from familiar tunes, such as *This is the way we tidy up, tidy up, tidy up* to the tune of *Here We Go Round the Mulberry Bush*. This is an example of using basic patterning skills to predict what will happen next: *I know that when we finish counting, we will tidy up*. This builds on all children's disposition to make connections: 'if' this happens, 'then' that will. Baron-Cohen (2020) argues that children with autism are often hyper-systemisers, who are particularly prone to notice *if-and-then* rules. Of course, this is part of a broader disposition in all of us to spot repeating patterns in events, so that we can predict, for instance, the timing of nightfall or the changes in the seasons, and avoid being taken unawares.

Patterns in song and stories

Songs and stories with a refrain or repetitive structure are very popular with the class. The repeating pattern encourages all children to join in and enjoy the communal experience of chanting and singing, as well as the shared anticipation:

> *You could almost feel as you got towards the end of the verse the expectancy in the room as the kids got ready to belt out, 'We all live in a Yellow Submarine!'*

A patterned structure gives access to a shared experience, which brings its own emotional satisfaction:

> *The value of repetition in a classroom for children with autism cannot be underestimated…They often gravitate towards the repeating chorus or the repeating line that signals the end of one part of the song before starting again with another verse. Michael absolutely loves 'Sleeping Bunnies'. But it is the action that he waits for. He watches the adult intently while they are singing and then just before you get to the part where you sing, 'Hop little bunnies, hop, hop, hop', he joins the other children and off they go. It's that predictability of pattern, that repetition, that makes it easy to follow and gives him the satisfaction and joy.*

However, for children with autism, a departure from the familiar version disrupts this predictability and can be upsetting: *Tom is a boy who absolutely loves this rhyme ('The wheels on the bus') but if he's listening to an edition that has different words, he becomes distressed*. Tom's reaction shows that he is paying attention to what is being sung and can identify the 'error' in the pattern, although he is not yet ready to accept 'editing'.

A child retelling a story from a picture book.

Traditional tales often have a patterned structure, which enable children to learn and retell them, with the support of a picture book. There are often opportunities for actions as well as chants, like running after the Gingerbread Man:

The refrain marks the beginning of the action, so as they sing 'Run, run, as fast as you can. You can't catch me, I'm the Gingerbread Man', off they run. Sometimes we find the children adding the refrain to their running games in the garden.

Providing props supports children in re-enacting stories, but sometimes the children organise their own. One group enacted *The three little pigs* by hiding from the child-as-wolf behind each of three large foam blocks, with the 'wolf' reciting *Little pig, little pig, let me come in* and the other children yelling, *Not by the hair of my chinny chin chin!* before running to the next block. They finally hid under the climbing area: *'and he couldn't get in there, so it was the perfect place for the brick house'.* The children are in effect enacting an ABCD pattern of actions and language, with the repeating unit consisting of: hiding, the 'wolf' coming and saying his line, the others responding, then running to a new hiding spot.

Another group enacted *Goldilocks and the three bears* by requesting bowls and spoons for 'porridge' at the water tray, then using the chairs in the book corner and finally moving to the beds in the home corner, each time following the language pattern:

They weren't sure of the words, but they were following the routine of 'First we do the porridge'. And one's going to be too cold, and one is going to be too hot. And the last one is always going to be just perfect, or just right.

Mary considered that in retelling these stories the children were beginning to think abstractly. They had identified the repeating unit of *too much, too little, just right*, which is applied to porridge, chairs and beds. They had analysed the patterned story structure, identifying which elements stayed constant and which changed and then adapting this to the contexts available in the nursery, such as the water tray. Other children re-enacted stories, using a range of resources: for instance, one child spontaneously retold the Billy Goats Gruff story, while 'tapping' shapes of sheep onto a baseboard. Until Mary pointed it out, we had not appreciated that children might be abstracting and generalising conceptual and language pattern structures in rhymes and stories by re-enacting them. Again, we should not be surprised, since we know that children readily spot and generalise language patterns, such as adding 's' to plurals like 'sheeps' or saying 'I swimmed', before they learn the exceptions to the rules. It is also obvious that songs, rhymes and stories which are part of an oral tradition have a repeating structure so that they are easy to memorise and join in with, as communal events.

Observing pattern in play

Routines and stories were aspects of pattern in the nursery unit which were adult-led. Mary also observed the children to find out how they were thinking and learning and identified patterns in their play.

... It's through play that we can see what children understand as they use play to practice ideas and concepts. And if we're really observing, we can see what the child is telling us that they understand and from that point we can plan the next steps.

Mary recounted the example of Jamie who initially presented as a much younger non-verbal child, with a very blank expression.

While waiting for his peers to arrive in the classroom Jamie began to explore the basket of bead strings. Initially he played with them in a sensory way but then I noticed him lining them up from longest to shortest in multiple configurations. After lining them up on the floor, he lined them up on the chair and then on the table, persisting even when they would roll off the table, by picking them up and doing it again. He took them all over the room to do the same thing over and over again. And he was genuinely excited by it. He'd lay the beads down and then he would have about three or four jumps up and down on the spot. And that to me was like, 'Oh, this is really exciting. I've done it.' And then he'd go back to picking them all up again and transporting them somewhere else and having another go. And I think what impressed me also was when he tried to do it on the table and the bead-strings were so long that they rolled off the other side of it. He didn't give up. He kept trying. He engaged in this play for about 40 minutes, totally absorbed and focused. And I suddenly thought, 'There's a real goal to what you're doing—... you are wanting to order them from longest to shortest. You have a rule'.

Jamie's beads arranged from longest to shortest

So, with that knowledge we can now plan for some similar types of play that we know he'll enjoy. But we'll use different materials and some of those might be very similar, like ribbon or rope or chain, and some of it might be about laying masking tape down on the floor. The more ways we can offer that same experience, the more likely it is to extend his understanding. And the other thing we can do is to name what he's doing. This time I didn't comment, I was just watching. I really wanted to see what the child was doing without interrupting him. But we will now introduce the language of maths into what he's doing. So that he's got the words there, even if he can only think of them to himself at this point in time, but it's beginning to put meaning in it for him and it helps him to understand his world better...He's one of those children that you look at, and you think, 'I don't know what your learning is going to be like ...' The problem is, [with a child with autism] you have to know what they know, so that you can build on it. Otherwise, they're just repeating the same thing over and over again, they're not building on that knowledge. And then, you know, they're kind of stuck.

In this way Mary used observation of a child's patterning in their play to inform her about their interests and thinking, which enabled her to provide further challenges to extend their learning. For other children, patterning might occur in radiating patterns, or symmetrical constructions. Mary also introduced repeating patterns, using small toys, pegboards, printing and a range of materials. However, unless she had observed Jamie ordering the bead-strings, she might not have recognised the level of thinking and determination of which he was capable. This was an important reminder to us, to allow time to observe children and provide opportunities for them to show us what they are interested in investigating.

Further challenges in ordering

Pattern experts

When invited to engage in the types of activities outlined in this book, some children in Mary's class quickly became pattern experts:

When we started the pattern project Richard really took on exactly what we were doing and seemed very focused about it and followed instructions incredibly well. So I can remember when Sophia (the teaching assistant) started him off and was playing alongside him and making a pattern on her board and showing him and explaining you know, this is 'red, blue, red, blue. We could say it's ABAB'. And within a few weeks he was really sophisticated in his ability to put patterns together, and it didn't seem to have been a huge learning curve. He just simply sort of thought 'Oh yeah, that makes perfect sense'. And he was soon doing his own patterns and making varieties of patterns.

Others took longer to respond but then showed sophisticated pattern understanding:

With Aisha you weren't hundred percent sure what she understood. It wasn't until one day she filled the pegboard with a pattern and brought it over to show us. At first, I'm looking at this pattern thinking 'I don't get it', but in the end, when she showed us, I can remember being just shocked that she had thought to do what she had done because it was a pattern but it required thinking outside the box, so much so, that I didn't immediately see it. Instead of going from left to right, she'd zig-zagged up and down the board. She could tell me what the rule was, so I knew that she understood. She had deliberately made a pattern and to have completed it can't have been easy.

Aisha's pattern

Interestingly, Aisha did not often have exchanges like this with Mary: it seemed that she was using patterning as a way of communicating with her.

These children showed the hyper-systematising predicted by Baron-Cohen (2020), but with the input on routines, patterns in stories and our patterning activities, they were able to use their systematising tendencies to support their mathematical development. Their experiences in the nursery prepared them for generalising pattern structures in more abstract ways:

Developmentally you can expect children to be able to take on abstract learning if they've had multiple experiences of re-enacting pattern and learning about pattern as we do with peg boards and sorting bears and so on. I think it's a very rounded experience and sets them up very well for understanding when a teacher talks about pattern.

In this chapter we have seen how one nursery teacher used pattern and repetition with children with autism and other speech and communication delays. From the very first days of the year, she made sure that routines and visual timetables helped children to predict what would happen next. She also used the repetition in songs, nursery rhymes and traditional stories to develop children's pattern knowledge, inviting them to participate in action and language patterns. With the appropriate opportunities and environment, she found that the children took ownership of these songs and stories and the patterns within them. Mary's participation in the pattern project made her recognise the mathematical patterns in the routines, in songs and stories and in the children's behaviour. Once she saw the children's repetitive behaviour as pattern, this made her value it more: she then added language to make the patterns more explicit and provided opportunities for children to generalise them.

Finally…

The example of Mary's class shows how everyday activities reveal pattern structures going beyond mathematics, for instance, with regard to time, in the sequence of daily events and prefaces to transitions. With rhymes and stories, children were not just identifying repeated actions and language, but abstracting pattern structures of the ideas involved, identifying *what stays the same and what changes*, and generalising these structures with different resources. Mary's children also show us how important pattern is providing order in many ways. The families' use of simple visual timetables transformed their 'lock-downed' lives and relationships. Children showed delight in spotting patterns and creating their own rules, even using patterns to communicate. It seems that patterns can provide satisfaction aesthetically, emotionally and socially, helping all children to overcome difficulties and build their learning and life skills.

Activities to try

- Observe children's patterning behaviour and identify a range of provision to encourage them to generalise pattern structures and rules.

- Introduce a new story with a repeating element or growing pattern, providing large and small scale props: observe how children use the pattern structure to take ownership and develop the story

- Consider how you might make explicit and develop the pattern element of daily routines, emphasising *what stays the same and what changes*

- dentify the sequence patterns of other activities, such as cooking, craft activities or outings, eg. *First we do this, then we do that, etc* and encourage children to predict what to do next.

.

8 The children's learning

Children now independently access pattern in their independent play much more than last year. Children have a new love of learning for maths.

A reception child's independent work completed during home learning

The impact of the project on the children's learning was wide ranging, including not only reasoning about patterns and numbers, but language, communication and social skills and other curriculum areas as well. As Mulligan and Mitchelmore (2009) suggested, children have very different levels of pattern awareness and the project teachers certainly found this to be true. Some children were happy just copying, completing and sometimes extending a pattern, while others very quickly began to identify and verbalise the unit of repeat, enthusiastically using abstract letter codes. We were all surprised by this sophisticated level of thinking – and then we wondered why.

A reception child creating an AB pattern with shapes and then choosing to record symbolically underneath

Perhaps, as adults, we associate letters with algebra and 'hard maths', which we think we should avoid teaching until children are older. While using letters in this way does not conform to the conventions of algebra (where *AB* would mean *A* multiplied by *B*) the use of letters may be seen as algebraic (or pre-algebraic) in the use of abstract symbols to represent pattern structures (Blanton et al. 2015, cited by Kieran et al. 2016). The evidence from the teachers contradicted the presumption that this was too hard for young children: the children loved talking about *AB* patterns and relished challenging themselves to create *ABC* or *ABB* patterns. Once the teachers got over their initial surprise, they too began using this language confidently and frequently. The language of pattern very quickly became the language of mathematics.

Reception children quickly responded to the regular pattern input. They began to talk about the patterns they observed in the environment, discussing what they observed and noticed. They began making patterns and very quickly picked up on the language we were modelling: talking about units of repeat, describing their patterns 'algebraically', and challenging their friends to guess covered parts or to fix patterns.

Examples of independent pattern exploration

The quality of children's language to express mathematical reasoning also improved: as one teacher commented: *Their explanations are so much more in depth*. An example of this was children's reasoning about the size of units which might 'work' to make a cyclical pattern, described in chapter 3.

ABC doesn't work. We can try AB first because that's two and it might work better.

This response was undoubtedly due to skilful teaching, as the teacher regularly asked children to predict, explain and offer a conjecture: however, it also shows the level of reasoning of which some young children are capable when taught in this way.

As well as pattern understanding, teachers found that the project had made an impact on children's number learning. This was due partly to the subitising activities. One nursery teacher commented:

The way they are thinking they are problem solving. With the subitising they are seeing the rearranging and the number is still five, so it is about conservation and all the ways five can look. So, they understand five better.

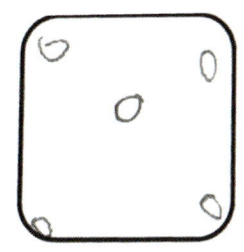

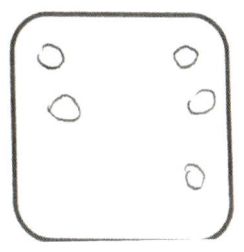

A reception child drew two ways to see five

Another teacher described the improved subitising of a reception child:

He could tell me immediately: 'That's five, that's three, I can see the pattern, it's like a little triangle' … and he's hanging his knowledge on that, and he's so proud of himself that he can.

Repeating patterns provided counting practice when children checked the size of the pattern units. For some number novices, this provided useful practice in counting small numbers: one reception teacher said: *Because she counted, she had a more secure knowledge of numbers.*

Counting while making a pattern

For higher attaining reception children, teachers noted an impact on their recognition of number patterns, such as adding one more and counting in twos, fives and tens. Children also reasoned about border patterns in terms of numbers: for instance, one child explained why a unit of five did not fit into 12 squares: *So: It's five and five and two more, so it's not enough and I need three more to make it.*

Will my border pattern work?

As mentioned previously, some children referred to units in terms of numbers: e.g. *So two works but I don't know about four yet.* This seems to show that patterning may prompt the beginnings of multiplicative thinking for some children. Some teachers developed this aspect by using units of repeat as measuring units. One child decided to measure the length of the table using an *ABCDEF* unit: he counted the units and said that he had used five and a half 'repeats'.

Children use units of repeat to measure the length of the table

The teacher brought him to another table (see previous photograph) where his friend had been measuring using an *ABB* unit of repeat. She asked him whether it would be the same number of repeats. He laughed and said: *No! I used six things in mine – ABCDEF, Neal used three – ABB, that's half of six so he will need lots more repeats.* He then counted them and said, *11 - Oh yes, that's because two lots of five and a half makes 11*.

We thought this was impressive reasoning for a five year old, seeming to show some insight into the inverse relationship between doubling and halving. At a simpler level, this activity shows the potential for patterning to help children to learn about unitising and to generalise, *the smaller the unit, the more you need*. It also shows the level of mathematical thinking and explanation of which young children are capable, when engaged in practical patterning enquiries.

The beginnings of an AB pattern outdoors

Children's progress

When the teachers repeated the four pattern assessment tasks at the end of the year, all children had made progress. While we might expect this, there were a few surprises too. Often, some children who had struggled to even copy or repeat a pattern at the beginning of the year were now using the unit of repeat to copy and extend the pattern confidently. In fact, children often surprised their teachers by going beyond the task set. One reception teacher reported: *I only asked if they could make an AB pattern, but they went on to make an ABC pattern and then explained it to me*.

An ABC pattern

Some teachers attributed improved results in end-of-reception assessments to pattern work. One school said it had increased children's scores for the Early Learning Goal (DfE, 2017) for *Shape space and measure* from 75% to 87%. They thought it had also contributed to a rise in *Number* scores from 74% to 80%.

A Masters student quantitatively analysed the assessment data from the London group for her dissertation (Wang, 2020). She found that the level of children's pattern awareness had a noticeable impact on the children's numerical understanding, particularly in the realms of counting objects, counting out a set of objects and early addition.

The impact seemed to permeate the wider curriculum: teachers noticed that other skills were improving which they believed were certainly linked to the work on pattern. One reception teacher thought that the pattern work had impacted on all 17 Early Learning Goals. Another said:

> *It's not just maths, it's language and it's PSED (Personal Social and Emotional Development) skills, because of the social skills that they have to have.*

Identifying each other's patterns   *Trying to make the pattern fit*

Because the children often worked together, making their patterns jointly or discussing challenges, teachers found they became more collaborative: *It's getting them to bond and be less competitive.* In terms of attitudinal development, teachers also noted that many children's levels of concentration, resilience and perseverance increased when they were working on pattern activities.

It seems that these practical and social problem-solving activities are highly engaging and cognitively stimulating for many young children. Patterning engages children in a range of learning, and because it can be applied in areas such as music, movement and outdoor learning, its potential impact is considerable.

Children are very creative in finding opportunities to explore pattern

Impact on families

One of the enjoyable things about conducting projects like this is the unexpected. In the beginning we were so caught up in the enthusiasm and excitement of both teachers and children, we almost overlooked the growing pattern community: parents and siblings.

Parents always comment how much their child loves pattern and are always creating patterns - even in their dinners.

While all of the teachers had been regularly including pattern activities in their 'stay and play' sessions or home tasks, it was really during the Covid-19 pandemic when we began to realise the impact patterning was having on home lives. Parents genuinely wanted to carry on with the pattern work their children had been exposed to in the classroom. A lovely, but unexpected, outcome of this was that often the younger siblings would take part in the pattern activities too. We have yet to see whether they arrive in school as competent pattern spotters already.

We have continued teaching patterns once a week via home learning and it has been brilliant, as the parents have shared so many fantastic photos and videos of the children creating these patterns.

Photograph sent in by a parent, whose reception child is absorbed in creating an AABB pattern on his kitchen floor.

Photograph sent in by a parent who recognised her reception daughter was writing her Father's Day Card using an AB Pattern.

One of the teachers created very detailed presentations for her children during the extended period of home learning. Realising that the parents would be supporting the children with this work, she always included a special slide 'For Grown Ups' which showed high expectations of them as well.

Exploring Pattern Teaching Order (FOR GROWN UPS)

- **Identify the unit of repeat** (or the pattern rule)
- **Continue the pattern** (for at least three more units of repeats)
- **Copy the pattern** (original pattern should contain several repeats, to ensure that the pattern unit is evident)
- **Edit**
- **Errors** (easier to spot an extra item, then a missing item or items swapped around)
- **Generalise** (use pattern experience/knowledge to create a pattern of the same structure but with a different medium)
- **Symbolise** (using symbols to record pattern eg. AB, ABB)

Example of the 'for grown-ups' slide

She reported back that many families were particularly pleased to receive their own pattern guidance. It therefore seemed that children were able to generalise their pattern understanding beyond the school setting, with beneficial effects on the mathematical dispositions of entire families.

Activities to try

Challenge children to:

- work together to create a pattern for others to continue or with deliberate mistakes for others to spot

- be creative in making the same pattern with different objects

- spot and describe patterns in different contexts

9 Reflections on teaching

I was a sceptic, we had used the compare bears, I thought this is the easy stuff, the pattern that they do in nursery. I've turned completely around, pattern is so embedded in every area, emotionally, socially, pattern is everything.

Teachers exploring and discussing pattern during one of the workshops

We were very lucky to work with such enthusiastic and committed teachers. Their creative implementation and adaptations of the pattern project was superb and enabled us all to reflect on how their teaching had developed: *We now know more about mathematics and also how to teach it better*. This chapter focuses on the impact the project has had upon us all, both from a subject and pedagogical point of view.

Teachers explore border patterns

An opportunity for professional development

ACME (2016) reports on the importance of teachers having opportunities to build on their knowledge of mathematics and how to teach it: professional development programmes that are on-going and involve collaboration are particularly powerful. We found this to be true. However, a key point about our pattern project was the lack of official guidance for teachers about pattern and the mathematics that was involved. For example, there was nothing about recognising the core unit of repeating patterns or naming this with letters in current English mathematics curricula or guidance. So, we were all building up our knowledge of pattern and how to teach it as the programme evolved. The collaborative element of the project was vital in allowing us to discuss what was happening in classrooms and what we might do next.

At each meeting we would focus on a particular type of pattern, which the teachers would try out in their own settings, to then report back at the next meeting. This format provided us with the opportunity to develop deeper knowledge about pattern, including how we might teach it and how children might respond to it.

A strong feature of each meeting was the conversation and collaboration that we all engaged in. The discussions covered a range of topics. We were always keen to show photos of children's patterns and recording, to talk about the children's responses and share evolving teaching strategies and resources.

Mine don't like border patterns.
Yes that's the same for mine. They do love linear patterns. I think they find turning the corner hard.
Some of mine just fill in the shape; they ignore the border completely.
I wonder whether we should have tried circular, before border?
Do you use lots of different equipment?
Cubes – yes, no distraction. But those little plastic people – no!
Using strips of paper [for repeating patterns] was revolutionary. Why didn't I think of that before?
The children are loving it. I am loving it.

A common topic was the development of children's language. The teachers were impressed by the way very young children were talking about their mathematical thinking: *They talk to each other about the patterns that they are making, what they are using*. A reception teaching assistant reported the improved speaking of three less confident children who enjoyed spotting errors and sharing their knowledge, for instance saying *No, no, no, this is how it goes*. Their confidence increased so much, that she taught them to make ABBC patterns and then *the three were teaching the whole class*. Another reception teaching assistant commented on the value of practical patterning for children with English as an additional language: *… because it's visual and they could watch you do it, they could follow what you did*.

As a consequence of this, all teaching staff were listening to the children more and were emboldened to use even more mathematical language. Not only was it allowing us all to notice more and observe children in different ways but it also prompted teachers to lead professional development, by sharing the project with staff in their own schools and more widely. This was not a common experience for many early years teachers, especially with teachers of older children.

Several teachers shared the project developments with other staff across the school and beyond, in the local authority and other schools.

One of the members of the pattern project led a staff meeting on pattern with her own teachers from early years to year 6.

Pattern pedagogy

One realisation was the power of deliberate mistakes to engage children in reasoning. Similarly, open-ended challenges fostered generalisation and abstract thinking. One reception teacher said: *It is just getting them thinking all the time. There is no right answer*. An unexpected consequence of the project was that all mathematics teaching became more open-ended and focused on children's responses. A reception teacher said that previously in the summer term:

I was more kind of 'teach' like year one, especially this time of year; now I've got into the mindset of just introduce and show and just put it out there and see what happens.

Many teachers remarked on how they listened to children more. A nursery teacher said that she would now *observe what they are doing, not just intervene, listen to their language, not talking all the time.* One reception teacher said she was now making sure activities and opportunities:

… are not so closed, they can explore, I'm not so much worried about the answer. Maths used to be prescriptive because there's such a lot to get through. Now I can be a more creative. It makes you listen to what they say, then using it to focus, to plan what next.

She added that she was teaching in this more open-ended way across the whole curriculum and listening to children more.

Teachers noticed the role of manipulatives as well as the collaborative nature of the activities, and the impact of this on learning:

Continue my pattern. That's really nice, doing it that way, it makes them talk to each other, it gives them confidence, they learn from each other. It's so concrete, they can touch it, feel it and look at it, its much easier for them to get their heads around. Because they can see if somebody else has created a pattern, or if there's a mistake or if there's something wrong, they can see that together.

Nursery children creating patterns together

The teachers realised that the children engaged well with pattern in continuous provision (which they said was often the main maths provision that the children used independently). This encouraged teachers to develop their outdoor maths provision.

Pattern activities as part of outdoors continuous provision

Teachers also became aware of the breadth of pattern linking with other aspects of mathematics. One reception teacher said she became aware of the importance of pattern providing: *A solid underpinning of number,* particularly in relation to subitising. We developed the range of classroom contexts and curriculum areas pattern could be taught through and linked with, including language, sounds, movements and time: *Pattern is now encompassed in all aspects of maths and literacy and all of the curriculum.*

Mathematics subject knowledge

We all found that the pattern project developed our mathematical subject knowledge and that this supported mathematical thinking. We realised that some of the most engaging activities were developing children's problem solving. Children did not automatically know if a particular pattern would work, so they needed to test it out; they needed resilience and perseverance to try and try again, before a solution was reached and teachers prompted them to reason or even record their thinking.

> *My subject knowledge about pattern has increased and I am now able to use the language and vocabulary of pattern confidently. I realise that we were not really teaching pattern before. Now we are able to support more children with reasoning and problem-solving skills.*

We were all surprised by the richness of pattern as a mathematical learning trajectory and the range of problem-solving activities which could be developed. Very quickly we realised there was more to pattern than printing red, blue, red, blue patterns or doing the butterfly paint symmetry activity. We asked questions that prompted children to analyse pattern structures, to generalise and to express these abstractly. We all learned that progression in patterning involved generalising complex pattern units and cyclical patterns as well as challenges such as spotting mistakes. We were excited that we could develop children's reasoning to a higher level than we previously thought possible: one reception teacher said: *These children are more than capable of going further, much further.* Another commented: *I can actually be part of that and try to create a thinking culture, you know, where learning is fun.* Teachers encouraged children to celebrate their achievements, for instance by taking photos of their work.

A reception child photographing her pattern work

Challenging assumptions

Engaging in depth with this new area of maths caused us all to reflect on previously held assumptions about what it means to be good at mathematics. One reception teacher said, *It's really interesting that children who don't present academically, do present within the pattern focus. I would have overlooked these children previously.* Elsewhere we have talked about children who we would have previously considered as number novices, and therefore mathematics novices, who then surprised us with their patterning. We noticed that the more the teachers engaged with the pattern project, the more adventurous they became in involving children to reasoning and generalising, for instance about complex structures. One reception teacher presented the image below and asked the children to describe it as a unit of repeat.

All said AABBC - identified as: shoe, shoe (AA), glove, glove (BB), duck (C).

After discussion they agreed it could be ABC as it was one pair of shoes (A), one pair of gloves (B) and one duck (C).

The teacher suggested ABCDE– each shoe is different (AB), the hands are different (CD) and then the duck (E).

The children would not accept this as an answer.

This not only showed the readiness of young children to discuss alternative descriptions of a unit of repeat, it also showed the teacher's high expectations in posing such a challenge to the children.

The work of these teachers also challenged often repeated assumptions about early years teachers' negative attitudes to mathematics. We were, and still are, very fortunate to work with a group of highly experienced committed and enthusiastic teachers:

I have loved the pattern project and something I will continue next year! I have learnt so much and the impact it has had on the children's learning has been amazing! I could talk for ages about it! My class is pattern crazy!

Pattern across the school

The pattern project across the school is working really well. I have done some monitoring this term. It was lovely when I was in a year 4 class and they were adding fractions so $\frac{1}{3} + \frac{1}{3} + \frac{1}{3}$ and so on, and the children were able to spot and identify a pattern emerging with the numbers.
(Mathematics subject leader and reception teacher)

The children were spotting that the denominators were the same (e.g. $\frac{1}{3}$, $\frac{2}{3}$, $\frac{3}{3}$, or one whole, $\frac{4}{3}$, $\frac{5}{3}$, $\frac{6}{3}$ or two wholes). These year 4 children had been exposed to pattern daily, and so noticed the repeating nature results? of the calculation. In this school all teachers from reception to year 6 began each mathematics lesson with a pattern starter. This was a result of teachers sharing the project work with colleagues.

Children from year 2 and key stage two year 4? involved in pattern activities

Teachers of older age groups were often impressed by the children's pattern work:

I was observed teaching pattern on Monday by our headteacher and primary lead for our Trust. They were really impressed with the children and could not believe what they were doing in Reception!

The headteacher said:

These children are using language, like 'unit of repeat', and I don't even know what that is. They are spotting patterns that I need to think about, and they are only in reception.

Teachers across one school responded positively to a session on pattern:

The CPD was very enlightening and encouraged me to look for opportunities to teach pattern discretely in maths lessons. The individual children's responses were interesting, highlighting strengths, weaknesses and misconceptions, also their knowledge, understanding and flexibility of thinking within pattern work. (year 3 teacher)

Before the pattern CPD, I wasn't really aware of the positive impact pattern work can have on children's mathematical understanding. It was great to see different examples of activities that can be incorporated into maths lessons and I am looking forward to continuing pattern teaching. (year 4 teacher)

Helen teaching seven year olds to recognise patterns when skip counting

There are many more lines of enquiry to follow with patterning which would be worth exploring.

Suggestions for further enquiry

How do children respond when patterns end with an incomplete unit of repeat?

Children tend to complete a linear pattern in one direction. Can they continue the pattern in both directions?

Present half patterns to complete with reflective symmetry

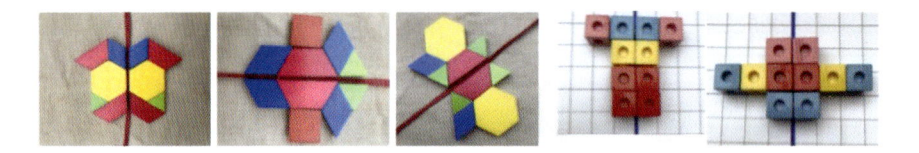

Quickly show a pattern, then hide it: can children remake (or draw) it from memory?

Can children spot, describe and create patterns in the environment?

References

Advisory Committee on Mathematics Education (ACME) (2016). *Professional Learning for all Teachers of Mathematics,* London: ACME.

Baron- Cohen (2020). *The Pattern Seekers: A new theory of human invention*. Dublin, Ireland: Penguin Random House.

Baron-Cohen, S., Ashwin, E., Ashwin, C., Tavassoli, T., & Chakrabarti, B. (2009). Talent in autism: hyper-systemizing, hyper-attention to detail and sensory sensitivity, *Philosophical Transactions of the Royal Society, 364,* 1377-1383.

Blanton, M., Brizuela, B. M., Gardiner, A. M., Sawrey, K., & Newman-Owens, A. (2015). A learning trajectory in 6-year-olds' thinking about generalizing functional relationships, *Journal for Research in Mathematics Education, 46* (5), 511-558.

Bragman, R., & Hardy, R.C. (1982). The relationship between arithmetic and reading achievement and visual pattern recognition in first grade children, *The Alberta Journal of Educational Research,* 28 (1), 44-50.

Clements, D. H. (1999). Subitizing: What Is It? Why Teach It? *Teaching Children Mathematics, 5* (7), 400-405.

Cuoco A., Goldenberg E. P. and Mark J. (1996). Habits of Mind: An Organizing Principle for Mathematics Curricula, *Journal of Mathematical Behavior*, *15,* 375-402. Cuoco_etal-1996.pdf

Department for Education (2013). *The national curriculum in England: key stages 1 and 2*. https://www.gov.uk/national-curriculum

Department for Education (2017). *Statutory framework for the early years foundation stage.* https://www.gov.uk/government/publications/early-years-foundation-stage-framework--2

Department for Education (2021). *Early Years Foundation Stage Statutory Framework*

Department for Education (2020). *Development Matters: Non-statutory curriculum guidance for the EYFS.*

Early Education (2021). *Birth to 5 Matters.* https://www.birthto5matters.org.uk/

Fyfe, E. R., Rittle-Johnson, B. & McNeil, N. M. (2015). Easy as ABCABD: Abstract Language Facilitates Performance on a Concrete patterning Task. *Child Development, 86*(3), 927 – 935.

Gifford, S. (2014). A good foundation for five-year-olds? An evaluation of the English Early Learning 'Numbers' Goal in the light of research, *Research in Mathematics Education, 16*(3), 219-233.

Griffiths, R., Back, J. & Gifford, S. (2016). *Making numbers: using manipulatives to teach arithmetic.* Oxford: Oxford University Press.

Gura, P. (1992). *Exploring learning: young children and blockplay.* London: Paul Chapman Publishing.

Helenius, O. (2018). Explicating professional modes of action for teaching preschool mathematics, *Research in Mathematics Education, 20*(2), 183-199.

Kidd, J. K., Pasnak, R., Gadzichowski, K. M., Gallington, D. A., McKnight, P., Boyer, C. E., & Carlson, A. (2014). Instructing first-grade children on patterning improves reading and mathematics. *Early Education & Development, 25*, 134–151. https://doi.org/10.1080/10409289.2013.794448

Kieran, C., Pang, JS., Schifter, D., & Ng, S.F. (2016). *Early algebra ICME-13 topical surveys.* Springer Open. http://link.springer.com/book/10.1007%2F978-3-319-32258-2

Marton, F. & Neuman, D. (1990). Constructivism, phenomenology and the origin of arithmetic skills. In L.P. Steffe & T. Wood (Eds.), *Transforming children's mathematics education: international perspectives* (pp.62- 75). New Jersey: Lawrence Earlbaum Associates.

Mulligan, J., & Mitchelmore, M. (2009). Awareness of pattern and structure in early mathematical development. *Mathematics Education Research Journal, 21,* 33–49. http://doi.org/10.1007/bf03217544

Mulligan, J., & Mitchelmore, M. (2016). *Pattern and Structure Mathematics Awareness Program (PASMAP): Book one - Foundation and Year 1*. Camberwell, Victoria: Australian Council for Educational Research.

Mulligan, J. T., & Mitchelmore, M. C. (2016). *Pattern and structure mathematics awareness program (PASMAP): Books 1 and 2*. Camberwell, VIC: Australian Council for Education Research (ACER) Press.

Papic, M.M. (2013). Improving numeracy outcomes for young Australian indigenous children. In L. English & J. Mulligan (Eds.) *Reconceptualising Early Mathematics Learning* (pp. 253-282). Springer Science & Business Media.

Papic, M., & Mulligan, J. (2007). The growth of early mathematical patterning: an intervention study, In J. Watson & K. Beswick (Eds), *Mathematics: Essential Research, Essential Practice — Volume 2, Proceedings of the 30th annual conference of the Mathematics Education Research Group of Australasia* (pp. 591-600).

Papic, M., Mulligan, J., & Mitchelmore, M. (2011). Assessing the development of pre-schoolers' mathematical patterning. *Journal for Research in Mathematics Education, 42*(3), 237-268.

Rittle-Johnson, B., Fyfe, E.R., Hofer, K.G., & Farran, D.C. (2017). Early math trajectories: Low-income children's mathematics knowledge from ages 4 to 11, *Child Development,* 88(5), 1727-1742.

Sarama, J. & Clements, D.H. (2009). *Early Childhood Mathematics Education Research*. Abingdon: Routledge.

Sayers, J. (2015). Building Bridges—Making connections between counting and arithmetic: Subitising, *Primary Mathematics, 13,* 22-25.

Sayers, J., Andrews, P., & Boistrup, L.B. (2016). The role of conceptual subitising in the development of foundational number sense. In T. Meaney *et al.* (Eds.) *Mathematics Education in the Early Years* (pp. 371-394). Switzerland: Springer International Publishing Switzerland.

Skinner, C. (1997). *Board games for the nursery (Set A)*. BEAM.

Standards and Testing Agency (2014). *EYFS profile exemplification for the level of learning and development expected at the end of the EYFS Mathematics ELG12 – Shape, space and measures.*

Thouless, H., & Gifford, S. (2019). Dotty triangles. *For the Learning of Mathematics, 39 (2)*, 13-18.

Thouless, H., Hilton, C., & Webb, T. (In press). Counting. In Y.P. Xin, R. Tzur, and H. Thouless, (Eds.) *Enabling Mathematics Learning of Struggling Students: International Perspectives*. Springer.

Thouless, H., Lewis, S. & Gifford, S. (2019). Staircase Patterns. *Mathematics Teaching, 269,* 37-40.

Threlfall, J. (1999). Repeating pattern in the early primary years. In A. Orton (Ed.) *Pattern in the teaching and learning of mathematics* (pp. 18-30). Cassell: London, UK.

Wang, Z. (2020). *Explore the relationship between preschoolers' pattern awareness and mathematical understanding.* [Unpublished master's thesis]. UCL Institute of Education.

Woodham, L. (2013). *Pattern sniffing*. NRICH. https://nrich.maths.org/9968

Acknowledgements

Project development group:

Chesterton Primary School: Isobel Henderson; Libby Meyer; John Stanghorn

Christ Church CE Primary School: Kasia Crossingum; Ruth James; Karen Moses; Kati Sells

Eastwood Nursery School: Ioanna Anagnostopoul; Mary Edgar; Karen Pearson; Evelyn McGeary

Granton Primary School: Sophie Grant; Katie McCarthy; Perishma Patel; Helen Wilczek

Ravenstone Primary School: Jack Davidson; Simon Lewis

Riversdale Primary School: Helen Barnard; Tracey Tattersall

Wandsworth Early Years Team: Catherine Gibson

Trialling group:

Attleborough Primary School: Jessica Skinner

Coltishall Primary School: Gemma Hudson

Coldfair Green Primary School: Mel Porter

Costessey Primary School: Jessica Blake

Dell Primary School: Sara Withall

Elm Tree Primary School: Fern Ross-Saunders

Holly Meadows Primary School: Rebecca Littlewood

Long Melford Primary School: Victoria Graham

Nelson Infant School: Ratna Sharma

St Clements Primary Academy School: Sian Dilley

Wells-Next-the-Sea Primary and Nursery School: Julia Norman

Additional contributors:

Sheena Preston, Paige Tynegate

Ofsted (2011). *Good practice in primary mathematics: evidence from 20 successful schools.* www.ofsted.gov.uk/resources/110140 .

The Power of Pattern
Patterning in the Early Years
Alison Borthwick, Sue Gifford, Helen Thouless

Pattern is fundamental to mathematics and should be at the heart of every early childhood mathematics curriculum. Pattern provides children with wonder, challenge, comfort and joy; establishing ways of thinking mathematically that provide a strong foundation for future learning. The 'Power of Pattern' helps practitioners to harness this potential by providing essential knowledge about pattern types and developmental progression that will shape their pedagogy. In a clear and accessible format, this ATM book showcases the remarkable patterning that young children are capable of. Drawing upon research in the field and their pattern project, the authors have provided guidance which is grounded in authentic early childhood practice with photographs and examples to inform and inspire every early childhood practitioner.

Dr Catherine Gripton, School of Education, University of Nottingham, UK

It is not often that a book lands in our laps which manages to blend research with practice so effectively and engagingly. This ATM book, on how mathematical pattern awareness is linked to wider numerical understanding, from Gifford, Borthwick and Thouless, is certainly one.
A report on the authors' research project, based on international research into the importance of pattern awareness to later mathematical learning, reveals much to think about for adults who work with children between the ages of 3 – 8, as well as for those working with older children who need re-engaging with mathematics. The book is filled with carefully developed tasks, together with thoughtful analysis of what we might observe and how to support children's development.
A must-have on the shelf of every one of us teaching mathematics.

Helen Williams

Packed with practical and intriguing activities, this beautiful book shows how to help young children develop their pattern awareness across a huge range of contexts. The research-based tasks build on children's natural interests and encourage them to work creatively together to analyse and produce interesting patterns and talk about their features. This book will help young mathematicians become great 'pattern sniffers' – one of Cuoco, Goldenberg and Mark's (1996) eight mathematical 'habits of mind'.

Colin Foster

Association of Teachers of Mathematics
2A Vernon Street
Vernon House
Derby
DE1 1FR

© Copyright

The whole of this book is subject
to copyright.

All rights reserved

Published August 2021

Printed in England

The Power of Pattern
Patterning in the Early Years
ISBN 978-1-912185-27-6

Further copies of this book may be
purchased from the above address

www.atm.org.uk